T0289974

MORE PRAISE FOR
Making Sense of Chaos

"This is the book I've been waiting for. It reinvents economics in terms of computerized agent-based modeling and complex adaptive systems theory developed over four decades. In many situations, 'complexity economics' is more predictive than classical economics and therefore provides a better basis for public policy. Doyne Farmer's life work with Santa Fe Institute is embodied here."

Stewart Brand, founder of the Long Now Foundation
and editor of *The Whole Earth Catalog*

"It can seem daunting to try to understand the globally connected, fast-moving, technology-driven, financialized economy of today; but there could be no better guide to this complex landscape than Doyne Farmer. This is a compelling book for anybody concerned about our economic future."

Dame Diane Coyle, author of *Cogs and Monsters*

"After 2008, everybody except conventional economists seemed to realise that conventional economics is failing us. I applaud this bold and exciting new approach, born out of the 21st century rather than the 19th. It's about time!"

Brian Eno

"In this riveting book, Doyne Farmer profoundly unravels the role played by complex systems in our economy. From the time he was living in a tent while working on climate models on the American west, to his more recent years in the hallowed halls of Oxford University, Farmer's lifelong journey is a testament to the creativity and perseverance needed to succeed in the rugged landscapes of multidisciplinary science."

César Hidalgo, author of *Why Information Grows*

"I can't think of a better person to help us make sense of chaos than one of the founders of the field of chaos theory, J. Doyne Farmer. A physicist by training, but with plenty of financial and economic street cred, Farmer takes on the formidable task of making complexity economics understandable, fascinating, and fun. And he succeeds!"

Andrew W. Lo, author of *Adaptive Markets*

"Doyne Farmer condenses a lifetime's worth of seminal contributions to chaos theory within this captivating book, and describes an ongoing paradigm shift within economics that can better address the problems facing our world by acknowledging and harnessing complexity. His genius is indisputable!"

Scott Page, author of *The Model Thinker*

"Standard economics has failed us when it comes to the challenge of climate change, and in this remarkable book Doyne Farmer explains why—and shows that a more complex understanding of how economies work yields insights that can help us see, and build, a workable future. Farmer's insights are more general, extending to all kinds of realms; indeed, it's exciting to sense the intellectual ground being broken here. But it's also of the highest practical importance; heeding his thoughtful counsel offers a path out of the box canyon where we're currently stuck as a species."

Bill McKibben, author of *The End of Nature*

"Our greatest challenges, such as climate change, force us to move beyond linear thinking to grapple with their complexities. Doyne Farmer's book shows us how to do so, bringing some clarity to the chaos. I gobbled it up—I hope it gets into the hands (and heads) of those who want to understand our most pressing problems and work towards solving them."

Hannah Ritchie, author of *Not the End of the World*

"Doyne Farmer is the world's leading thinker on technological change. For decades he has focused on the question of how we can make sense of the data of today to see where the world is going tomorrow. This wonderful book applies these insights to economics, addressing the big global issues of environmental sustainability, and the well-being and prosperity of people around the world."

Max Roser, Founder of Our World in Data

"This book is a real achievement from which I learned a great deal. The economics profession should be much more open to Farmer-type complexity approaches. I hope this book is an inspiration to young scholars from many disciplines concerned with economic questions."

Lawrence Summers, former US Secretary of the Treasury

"Mainstream economics achieved an impressive level of mathematical rigor, but at the expense of grossly simplified assumptions and an exclusive focus on the equilibrium, making it ineffective in helping us solve real-world problems. Is 'predictive economics' even possible? In this remarkable book, Doyne Farmer, a cofounder of the modern science of chaos, shows that it is. He convincingly argues that we need to build realistic agent-based models, grounded in data, that will enable us to find effective ways to run our economy and navigate societal challenges. A must-read."

Peter Turchin, author of *End Times: Elites,*
Counter-Elites, and the Path of Political Disintegration

Making Sense of Chaos

Making Sense of Chaos

A Better Economics for a Better World

J. DOYNE FARMER

Yale UNIVERSITY PRESS
NEW HAVEN AND LONDON

First published in 2024 in the United Kingdom by Allen Lane,
a division of Penguin Random House UK, and in the United States
by Yale University Press.

Yale University Press books may be purchased in quantity
for educational, business, or promotional use. For information,
please e-mail sales.press@yale.edu (U.S. office)
or sales@yaleup.co.uk (U.K. office).

Typeset in 12/14.75 pt Dante MT Std by Jouve (UK), Milton Keynes
Printed in the United States of America.

Library of Congress Control Number: 2024931693
ISBN 978-0-300-27377-9 (hardcover : alk. paper)

A catalogue record for this book is available from the British Library.

10 9 8 7 6 5 4 3 2

This book is dedicated to my childhood mentor,
Thomas Edward Ingerson (1938–2019).

Contents

Acknowledgments

During the ten years it took me to write this book, I often felt like a monkey with a typewriter; without extensive help from my friends, colleagues and editors, it would have been incomprehensible garbage. I should first thank my agent, John Brockman, who nagged me to write this book for thirty years, and my other agent, his son Max Brockman, who helped push it over the finish line. On the editorial front, I would like to thank Vince Bielski, Seth Ditchik, Edward Kastenmeier, Sienna Latham, Dorothy Nicholas and Andrew Weber for valuable suggestions. Kurt Vonnegut once said that when he wrote, he always imagined he was talking to his sister; Shanny Peer played that role for me, but much more: She encouraged me, forced me to be clear, made insightful suggestions and spent innumerable hours editing every line. Sara Lippincott did a superb edit from front to back, further streamlining my rambling thoughts. I was lucky to get Keith Mansfield as my editor at Penguin, who made creative, thoughtful suggestions and empowered me to make the book lively and fun. I was similarly lucky to have Alba Ziegler-Bailey as my copyeditor, who painstakingly read every word, improved my phrasing, added hundreds of needed commas and then generously let me take some of them out. Luca Mungo wrestled my motley collection of crude figures into a consistent and well-rendered form, and David Mackintosh produced a richly ironic visualization of the wilderness of bounded rationality. My fellow complexity economists Brian Arthur, Rob Axtell, Peter Barbrouck-Johnson, Jean-Philippe Bouchaud, Maria del Rio-Chanona, Alissa Kleinnijenhuis, François Lafond, James McNerney, Penny Mealy, Jose Moran, Scott Page, Marco Pangallo and Anton Pichler helped create the substantive material for the book, guided me in its articulation and gave important feedback. My open-minded, mostly mainstream colleagues John Geanakoplos, Jeremy Large, Andrew Lo, Larry Summers and Alex Teytelboym had the courage to engage with me, read the book and correct many errors (but those that remain are all mine). Meanwhile, my friends Sander Bais, Kingsmill Bond, Chris Magee, Fotini Markopoulou and Peter Turchin helped me debug my explanations to make them clear, and my friends Karen Lawrence, Steve Lawton and Robb Lentz even went above and beyond the call of duty and read the manuscript twice! (Giving very useful feedback each time.) My friend and musical partner Tim Palmer improved my description of how we predict the weather. I would particularly like to thank Eric Beinhocker and Cameron Hepburn at Oxford for their years of collaboration and support. Finally, the underlying research presented here would not have been possible without the people and organizations who have funded it through the years, with special mention for Jack Mezaros, Leslie Haroun, George Soros, Baillie Gifford and most of all Gerald Smith, whose generous no-strings-attached contribution allowed the complexity economics group at Oxford to thrive, and Dorothy Nicholas, whose organizational skills keep us from falling into chaos.

List of Illustrations

Prologue

There has always been a strand of thought [. . .] that holds that we cannot hope to understand the major events in the life of an economy, and perhaps also its everyday behavior, without entertaining principles of disequilibrium.

Edmund S. Phelps (1991)[1]

On March 16, 2020, I mobilized a team of some of my best Oxford students, postdocs and former-students, and we began a crash program to build a model to predict how the economy would respond to the COVID-19 pandemic. It was clearly only a matter of time before the disease would become widespread and many governments would instigate lockdowns, with an enormous economic impact. But there are many ways to implement public health measures. Governments would need to decide what restrictions to impose and which industries should be closed. Was there a sweet spot that might provide a reasonable compromise between preventing the suffering that would be caused by infection and mitigating the suffering from economic stress? Could we predict the economic impact of COVID? (At the risk of spoiling the punchline, the answer is yes.)

COVID was going to shock the economy from both sides. On the demand side, with or without government-imposed lockdowns, consumers would curtail many of their former activities; on the supply side, non-essential industries where workers were in danger of infection were likely to be forced to cut back or shut down completely. To understand the economic impact, we needed to forecast the shocks in advance and understand how they might be amplified as they propagated through the economy.

Our first task was predicting the shocks. To do this on the demand side, we used a study compiled by the US Congressional Budget Office in 2006, which predicted what would happen to demand for the products of different industries during a flu epidemic.[2] Although the study took the SARS flu virus as the disease template, and there are some differences between SARS and COVID, it provided a reasonable guess at how consumers would respond. As it turns out, they got a few things wrong, such as anticipating that demand for health care would go up (it went down, because of deferral of routine treatments), but in general their predictions were sound.

Predicting the supply shocks was harder, and we had nothing to guide us. Two remarkable scientists and former graduate students of mine, Penny Mealy and Maria del Rio-Chanona, led the effort to assemble this piece of the puzzle. When the pandemic broke out, Penny had already left Oxford and returned to her hometown of Melbourne, Australia, where she was a postdoctoral fellow at Monash University. She had to be awake late at night to make our daily Zoom calls, but this arrangement had the advantage that she and Maria could hand problems back and forth, so someone was working on the model twenty-four hours a day.

Maria and Penny took advantage of a remarkable data set compiled by the US Department of Labor that goes by the acronym O*NET, and provides detailed information about almost 800 different occupations.[3] It includes key facts we needed, such as the typical proximity of workers to each other, which we used to estimate which kinds of workers would be exposed to infection. It also provided information about the number of employees who work in each industry. O*NET enabled Maria and Penny to construct a remote-labor index that predicted which occupations would be able to work from home.

To help us understand which industries governments might shut down, another of my former students, Marco Pangallo (who you will hear a lot about later), found an Italian government list of non-essential industries.[4] The combination of Maria and Penny's remote-labor index and Marco's list of non-essential industries allowed us to predict the drop in the labor force in each industry. By assuming that output would drop proportionally to the number of workers removed from the labor force, we were able to predict the direct economic impact the pandemic shutdown would have on each industry.

Our paper predicting the supply and demand shocks for the second quarter of 2020 was publicly released one month later, on April 14, 2020.[5] For reasons of data convenience, our paper only addressed the US economy.[6] Compared to the pre-COVID period, we predicted that the direct shocks would reduce US GDP by about 20 per cent, jeopardize 23 per cent of jobs, and lower total wage income by 16 per cent. (These numbers are quarterly, so if the shock lasted for only three months, the annual effects would be proportionally smaller.) We also predicted that there would be a relatively small effect on high-wage occupations (a 6 per cent job loss) and a large effect (41 per cent) on low-wage occupations. Low-income workers also had a much higher probability of becoming infected.

In hindsight, these predictions were broadly right. Nonetheless, we knew they were missing a key effect that would make the impact even bigger: These were only the *direct* shocks, which would be amplified as they reverberated around the economy. Understanding how this would unfold is complicated because supply and demand shocks flow in opposite directions. Goods and services flow downstream from supply to demand, whereas demand shocks flow upstream: If the demand for a consumer product – say, bicycles – drops, the demand for bicycle tires drops, which means the demand for synthetic rubber drops. Conversely, supply shocks move downstream: If there's a shortage of workers to make synthetic rubber, this creates a shortage of tires, which creates a shortage of bicycles. Supply and demand shocks interact as they move downstream and upstream, colliding with one another and amplifying the initial shocks. This can reduce economic output even further.

The economic impact of the COVID pandemic combined three exceptional features that made standard models inadequate: First, the shocks were highly industry-specific, making it essential to model the economy with a high degree of resolution; second, the shocks affected supply and demand simultaneously; and, third, the shocks were huge and happened really fast.

Standard economic models assume equilibrium, which means that supply and demand are always equal. The COVID shocks hit the economy so hard and so quickly that they knocked it far out of equilibrium, making standard models unsuitable. We needed a model that could track the dynamics of the shocks without assuming equilibrium, and we needed to build it very quickly. The fact that there were no well-recorded historical examples of similar pandemics made this particularly challenging.

Because we were based in England, and because we knew the British government was actively considering its options, we decided to shift our focus to the UK. The effort to build our model of COVID shock dynamics was led by my Austrian graduate student Anton (Toni) Pichler, who had just completed his DPhil (what Oxford calls a PhD) in mathematics. We needed to track the factors that would lower economic output and understand their impact. On the demand side, consumers might reject a good or service for fear of infection; for example, even though airlines and other modes of transportation were deemed essential, their customers largely abandoned them. On the supply side, an industry might not have the workers and material inputs needed to produce its product, or it might be deemed inessential and be forced to shut down.

To understand how all these factors would affect each industry we needed what is called a *production function* in economics. A production function is a kind of recipe that tells how much can be produced with a given amount of each input. Making steel, for example, requires iron, coke, coal and labor, all in the right proportion. Unfortunately, there was no standard economic model of the production function that was sufficiently well grounded in data to be useful. The key issue was substitutions: Which inputs could be substituted, and to what extent? We needed to know this industry by industry. The steel industry also has inputs from the management consultant industry and the restaurant industry (to run cafeterias), but it can get by without these if it needs to. Neither of the standard choices for production functions worked – either they allowed too many substitutions or none at all. We needed something more realistic. Luckily, we managed to convince a company called IHS Markit to perform a survey of their industry analysts *pro bono*, which they did in only two weeks, identifying the critical inputs for each of fifty-two different industries.[7]

We modeled each industry as if it were a single firm; this is a crude approximation, but it was good enough for our purposes. We assumed that each industry began the pandemic with a normal level of inventory. As problems occurred, inventories dropped; if an industry ran low on a critical input, its output dropped. Similarly, if the demand for the industry's output dropped, or if it didn't have enough workers because of COVID restrictions, its output would drop. Bear in mind that one industry's outputs are often another industry's inputs. Our model tracked the demand shocks as they flowed upstream and the supply shocks as they flowed downstream, colliding with one another and lowering the overall output of the economy. It also allowed us to feed in the shocks we predicted in our first paper (adjusted for the UK), watch them propagate, and forecast the overall hit to Britain's GDP, unemployment and other economic indicators under different lockdown scenarios.

In early May 2020, the UK had already been in a severe lockdown for more than a month, and the government was considering what to do next. We ran our model for several different scenarios and evaluated their economic impact. For each scenario, we also estimated the expected number of COVID infections, using standard methods from epidemiology. Extending the severe lockdown would reduce the number of infections but increase the economic hit, while relaxing it entirely would cause a lot of deaths, but with less economic stress. There were no good choices, but we found a 'least bad' choice, leaving all upstream industries open while banning person-to-person participation in consumer-facing industries. Under this scenario, infections were only a little worse than under a total lockdown, but the economy suffered much less. On May 8, completely exhausted from our marathon effort, we sent our paper to UK government officials.[8] A few days later they relaxed the lockdown along the lines we recommended.

A post-mortem analysis shows that we did extremely well. We predicted a 21.5 per cent contraction of GDP in the UK economy in the second quarter, with respect to the last quarter of 2019, which was remarkably close to the actual contraction of 22.1 per cent. To put this in perspective, the median forecast by several institutions and financial firms was 16.6 per cent and the forecast by the Bank of England was 30 per cent. Of course, it's easy to get lucky forecasting a single event, but, because we had made forecasts for every industry sector, we actually made fifty-two forecasts, and most of those were pretty good too. Our post-mortem analysis confirmed that we succeeded because our model had what I call *verisimilitude* – it rendered the most important mechanisms affecting the economy with enough realism to provide accurate predictions. Our model has since been used by another group, who (retrospectively) used it to predict the shocks to the Belgian economy, where it performed similarly well, and it has been used to forecast the consequences of the Ukraine War for the Austrian economy.[9]

Our effort provides a proof of principle for the usefulness of complexity economics, which is the topic of this book. Complexity economics is an interdisciplinary movement of rebellious economists and other scientists who aim to better understand the economy using principles that are completely different than those of standard economics. Our COVID model was the first to use complexity-economics methods to make an accurate prediction of an important economic event, outperforming the standard models. This was done in real-time – meaning before it happened – so no one can accuse us of cheating. Complexity economics is a movement that has been a long time coming, but it is now taking off – complementing and competing with the standard methods that have dominated mainstream economics for more than a century.

In my career I have had the good fortune to participate in three major scientific revolutions in the understanding of chaos, complex systems and machine learning. Each of these movements began on the fringe of the scientific establishment and slowly – then suddenly – moved into the mainstream. Complexity economics, which applies complex-systems ideas to economics, is a revolution-in-progress whose moment is about to come.

Here I explain what complexity economics is, what it has already done, why its time is ripe and how it can make the world a better place. This is a story of scientific discovery, which is the best adventure available in the modern world.[10] This book is simultaneously a reflection on what the economy is, an account of how complexity economics can better guide it, a chronicle of a scientific revolution, a story about the evolution of science and the scientific method, and a memoir. I hope you enjoy it.

Introduction

Being an optimist, I have hope that the profession [of economics] will someday get into a scientific mode, based on anthropological observation of behavior and computer simulation of interaction.

Barbara Bergmann, pioneer of micro-simulation (1980)[1]

We live in an age of increasing complexity – an era of accelerating technology and global interconnection that arguably holds more promise, and more peril, than any other in human history. New means of flourishing lie tantalizingly at our fingertips, but Armageddon is all too easy to imagine. The fossil fuels that power the global economy have generated remarkable wealth, but they threaten to destroy it through climate change, an immense tsunami rolling toward us from the future. Automation and artificial intelligence are creating prosperity for some and unemployment for others. Financial crises and growing inequality have contributed to global polarization and the retreat of democracy.

Many of these problems are rooted in the economy.[2] Climate change is caused by economic activity, automation is intended to increase economic productivity, and financial crises and inequality are the result of poor management of the economy. Public policy-making for guiding the economy relies on theories and models that frame government debates, refine options, and provide arguments for and against proposed changes. Just as important, economic models and theories shape our intuition and affect how we think about the world and the decisions we make.

Unfortunately, the guidance provided by existing economic models has often failed. The 2008 global economic crisis offers a good example. A couple of years before the crisis started, top economists at the New York Federal Reserve Bank became concerned about the possible dangers of a crash in the housing market. They asked the Federal Reserve Bank US model (FRB/US), one of the most well-developed economic models in the world, what would happen to the national economy if the average house price dropped by 20 per cent.[3] 'Not much,' was the answer.[4]

The economists' concerns were remarkably prescient: Between January 2007 and January 2009, the average price of a house sold in the US dropped by about 23 per cent. The FRB/US model's prediction had been horribly wrong, underestimating the effect on the US economy by a factor of 20. At the time, the financial crisis of 2008 was the worst period of global economic decline since the Great Depression of the 1930s. In the United States alone, five million people lost their jobs, unemployment rose to 10 per cent and the economy lost some $10 trillion – and the effect on the rest of the world was even bigger. The failure of economic models was universal. The President of the European Central Bank, Jean-Claude Trichet, later confessed to his colleagues, 'As a policymaker during the crisis, I found the available models of limited help. In fact, I would go further: In the face of the crisis, we felt abandoned by conventional tools.'[5]

Contemporary science and technology can help us understand and solve the world's problems in ways that differ dramatically from those offered by the 'conventional tools' of economics. We are now able to collect data on economic activity at an unprecedented scale. Modern computers are capable of modeling the global economy with a remarkable level of fidelity and detail, and could be used to guide us through crises and help create an economy that is sustainable, fair and prosperous.

This is not happening, at least within mainstream economics. There have been improvements to the conventional tools, but they don't go nearly far enough. The problems are inherent in the conceptual framework that dominates the profession. While this framework is useful for well-specified problems, like designing a more efficient auction or understanding the strategy of labor negotiations, it is ill-suited to addressing big, messy, complicated, real-world problems. The core assumptions of mainstream economics don't match reality, and the methods based on them don't scale well from small problems to big problems and are unable to take full advantage of the huge advances in data and technology. For important global issues, such as climate change, standard economic theory's wrong answers have provided a rationale for inaction, leading us to the brink of global catastrophe. Nobel Prizes have been given to economists whose theories were based on idealized arguments with no empirical support, providing fodder for neoliberal policies that led to extreme inequality and fueled sociopolitical polarization.

This book is not about criticizing economics – there are plenty of books doing that already. It is about *how to do economics differently* to take proper advantage of big data and computer power, to create economic models that make better predictions and to give better policy advice about the hard problems facing the world. To be clear, this book is mostly about macroeconomics – the branch of economics that focuses on large-scale phenomena like inflation and unemployment – and finance, which is about money and investments. We will focus exclusively on economic *theory*, rather than economic data analysis using statistical methods. This is called *econometrics* in economics and *machine learning* in the field of artificial intelligence. (In both cases, models are formed simply by fitting functions to data – economists tend to favor linear functions, while computer scientists prefer neural nets.)

By *theory*, I mean building models of the world based on assumptions about how it works. Newton's laws of gravity and motion, for example, make it possible to predict the motion of comets and planets and explain why the planets move as they do. Statistical methods, in contrast, illuminate regularities in data and exploit them for prediction without making underlying assumptions. Econometrics has begun to analyze large data sets and is currently yielding many important results, while machine learning is exploding. Nonetheless, though statistical analysis is extremely useful, there are fundamental limits to what it can do.[6] Statistical analysis can help us build a picture of the world as it is, but it can make inferences about the future *only if* the underlying behavior remains the same. In a genuinely new situation, statistical regularities may shift. A more reliable way to anticipate what will happen when the world changes is with a theory – as long as it is based on mechanisms that do not change. A famous example of why this is so important in economics is the Lucas critique, which I describe in Chapter 5.

Economics is fundamentally harder than physics because, unlike planets, people can think, and their behavior can change as a result. Economic theory has to explain the behavior of *agents*, like households or firms, who make decisions, such as what to consume, where to work, how much to work, where to invest, what to produce, how much to produce, what price to charge and how to innovate. There is a standard template for building economic theories that incorporate agents' ability to reason, which evolved over the last 150 years and gelled into its present form in the 1970s. This template, which I will call *standard economic theory*, assigns each agent a utility function describing her preferences. For example, in standard macroeconomic models, households get utility from consumption and firms get utility from making profits while minimizing risk. Under standard economic theory, each agent makes decisions that make her own utility as large as possible. To find these decisions and understand their economic consequences, economists write down and solve equations that express this in mathematical terms. This is the basis of all economic theories that are taught in textbooks, and all the theoretical models that economists use to evaluate new policies and provide guidance to the economy.

The science of *complexity economics* offers a completely different alternative. Complexity economics uses ideas and methods from the science of complex systems, an interdisciplinary movement I helped found in the 1980s that spans the physical, biological and social sciences.[7] Complex systems is the study of *emergent phenomena*. These occur when the behavior of a system as a whole is qualitatively different from that of its individual parts. A good example is the brain: An individual neuron is a relatively simple device that takes in stimuli from other neurons and produces new stimuli. An individual neuron is not conscious, but somehow the 85 billion neurons in the human brain work together to produce consciousness and thought.

The economy is a wonderful example of a complex system, with many different emergent phenomena. The economy allows us to support each other by specializing – concentrating our effort on what we do best, while letting others do what they do best. By coordinating our efforts, it allows us to build things like rocket ships or laptops that no individual could ever build on her own.

While standard economic theory can sometimes provide valuable insight about emergence, this is not its strength – it was built for other purposes. In contrast, complex-systems scientists have identified and developed a standard toolkit of concepts and methods that they use to understand emergent phenomena such as brains, ant colonies, the internet or cities. Complexity economists like me are rebels who have adapted the complex-systems toolkit to study economics. I'm part of a band of some fifty-or-so complexity economists worldwide (the term was coined by Stanford economist W. Brian Arthur, one of its pioneers).[8] Complexity economists are a loose-knit group who often have divergent views. Some aspects of the perspective I present in this book are held in common; many are my own.

Complexity economics is completely different from standard economic theory. To begin, standard economic theory and complexity economics use very different models for how agents reason and make decisions. Standard economics usually assumes forward-looking agents who are intelligent enough to reason their way through any problem. These hypothetical rational agents are a bit like *Star Trek*'s Mr Spock: They use all available information to make the best possible decision – the one that yields highest utility. While the now-mainstream field of behavioral economics draws on facts from psychology and sociology to challenge this view, derisively referring to rational agents as '*Homo economicus*', there is still no widely used workable substitute. Although behavioral economics is now almost universally accepted, the economic models that continue to influence public policy are still largely based on *Homo economicus*. This is because it is hard to incorporate the way real people reason into standard economic theory. The elements of standard economic theory were designed to work together as a whole, and when some are modified the others become inappropriate. Creating economic models that effectively incorporate behavioral realism to make useful predictions may be the most important problem in economics today.

Complexity economics, conversely, assumes from the outset that agents are *boundedly rational*, meaning they make imperfect decisions and have limited ability to reason. Our models for boundedly rational agents are less prescribed – they can be based on psychology experiments, census data, expert opinions or a variety of other sources (see Chapter 6). Agents can learn to achieve goals, but they typically only partially achieve them.

The tools of the trade are very different. Standard economics models are written in terms of equations, and standard economists solve them by hand, if they can, or with computers if they can't. Though we use equations too, computer simulation is the workhorse of complexity economics. Digital technology enables us to create analogs of the real world inside the computer, sometimes called *digital twins*. Essential events in the real world have their counterparts in a simulated world. Scientists now use computer simulations to study just about everything – galaxy formation, protein-folding, the brain, epidemics, battle tactics and traffic jams. While mainstream economists use computers to solve equations, this is not simulation in the usual sense because there is no attempt to create analogs for how things actually happen in the real world.

The simulations used in complexity economics are called *agent-based models*.[9] As the name emphasizes, the individual building blocks of these models – such as households, firms or governments – make decisions; they have agency.[10] Algorithms describing how each agent makes decisions are programmed into the computer. As the agents gather information and make their decisions, their actions change the state of the economy, which in turn generates new information, causing the agents to make new decisions, and so on. The simulations can be repeated indefinitely, to study the interactions of individual agents over time. In agent-based models, economic phenomena emerge from the bottom up, just as they do in the real world. The algorithms programmed into the computer to model decision-making might consist of simple behavioral rules that social scientists have uncovered, like 'imitate your neighbors' or obvious rules of thumb that everyone follows, like 'buy undervalued assets'. Or they might be learning algorithms that capture how agents adapt their decisions over time to improve their effectiveness. Computers can easily deal with complicated, messy facts, enabling us to model decision-making with as much realism as we wish. And there are additional benefits of this realism: In the presence of uncertainty, simple rules of thumb can be surprisingly effective – sometimes more effective than complicated reasoning.[11]

There are many advantages to using simulations rather than mathematical equations, but one of the most important is the ability to model the diversity of the actors in the world – what economists call *heterogeneity*.[12] Modeling heterogeneity is a hot topic in mainstream macroeconomics, but it is very hard to fit this into the standard framework. Computers, in contrast, easily keep track of diverse agents, making it possible to do this as realistically as needed.

The theoretical frameworks of standard economics and complexity economics are also very different. Standard economics draws heavily on mathematical methods from the theory of optimization. This is like finding the highest peak in a landscape (except that here it is an abstract landscape representing utility). Standard economists like to prove theorems about their models whenever they can. We complexity economists also use mathematics, but usually do so to understand something observed in simulations. The mathematical tools are taken from a much broader palette, using methods and concepts from fields as diverse as dynamical systems, statistical physics, ecology and evolutionary biology.

Last but not least, standard economic theory assumes from the outset that transactions only take place when supply equals demand, a state that is called *equilibrium*. Complexity-economics models, in contrast, do not necessarily assume equilibrium, but regard it as an emergent property when it happens.

Taken together, all these elements lead to a very different way of dealing with change. Change in the economy can come from without: Think of the COVID pandemic, which was an outside force affecting the economy. Change can also come from within, like the financial crisis of 2008, which arose via the economy acting on itself. Standard economics has a difficult time dealing with changes that arise from within the economy. The leading mainstream models of the financial crisis, for example, have to postulate the existence of outside 'shocks' to generate the necessary change.[13] It is obvious that is not how it really happened, but standard models can only explain it 'as if' it happened this way. In contrast, agent-based models are inherently dynamical, making it natural to explain how the economy changes from within. We will see examples in Chapters 3, 7 and 11 illustrating how complexity economics replaces 'as-if' reasoning with 'as-is' reasoning, providing more sensible explanations for how the economy changes from within itself.

Perhaps the biggest advantage of complexity economics is its ability to solve hard problems. A model is *tractable* if it is easy to build and use to answer questions. Standard economic theory yields tractable models in simple settings, but this breaks down as things get complicated – it becomes too hard to solve the equations, and it becomes increasingly difficult to add new features to a pre-existing model. As a result, when a problem gets complicated, mainstream economists are forced to oversimplify by leaving things out.

The financial crisis of 2008 is a good example where this leads to problems.[14] All the models at the time ignored the possibility that homeowners and businesses might default on their loans.

Economists knew that default was important, but including it in their models was difficult, so they left it out. As it happened, default played a central role in the crisis. The traditional models, updated since the crisis, now include default, but only in a simplified fashion – the revised models still omit many essential economic factors that may well figure in future crises.[15] There is no model based on standard economic theory that comes close to including all the facets of the economy that could potentially lead us into trouble again.[16]

Making realistic models requires good data. Agent-based models are naturally suited to make use of the vast quantities of data currently available. Worldwide, tens of millions of firms and billions of households make trillions of transactions annually. It would be extraordinarily useful to build a detailed map of the global economy, but, remarkably, no such map exists. If we built one, it would show us the structure of the economy in detail and enable us to track its changes over time. We are now developing agent-based economic models at the level of individual firms and households that can use such data. These models study the economy from the bottom up: Macroeconomics emerges from microeconomics.

The choice between standard economic theory and agent-based models reflects a broader debate in social science. Milton Friedman, a prominent twentieth-century economist, argued that models should be evaluated *solely* on how good their predictions are, rather than on the plausibility of their assumptions.[17] Even if the assumptions of a model are far-fetched, he wrote, the economy may behave 'as if' they are true. He used the example of expert billiards players, who may not understand the laws of physics but play billiards as if they did. Behavioral economics has made it abundantly clear that people don't make decisions in the way that standard economic models assume they do. Nonetheless, to justify their models, economic theorists continue to assert that the economy behaves 'as if' people did.

There is universal agreement that models should be judged based on their ability to predict empirical facts. The debate is whether this should be the only criterion. In social science, models often fit the facts loosely and no model fits them tightly. We should be very suspicious of models that require implausible 'as-if' arguments. Instead, I believe we should follow what I call the *principle of verisimilitude*: Models should fit the facts *and* their assumptions should be plausible. Assumptions that seem wrong from the outset are more likely to lead to false conclusions than plausible assumptions. We need to replace 'as-if' reasoning with 'as-is' reasoning.

The principle of verisimilitude recognizes that models need to contain the key features of the phenomena they attempt to explain, but they needn't be literal representations of the world; models are abstractions, and we don't need to capture every detail. When we talk about building simulations, we aren't trying to create *The Matrix*. *Verisimilitude* just means capturing the essential components as realistically as we can. Good models should be as simple as possible, but no simpler. Agent-based models can be complicated or simple; computers easily keep track of details that are complicated to state mathematically, so we can include as many features as we need to. Most important, we can easily add new features without changing existing features, incrementally increasing the verisimilitude of our models of the world.

There is more to complexity economics than just simulations. As already noted, it draws on conceptual frameworks that have been developed in other fields. We will see examples where it is useful to start by modeling economic agents 'as if' they make decisions at random, and then add enough intelligence to make a model with more verisimilitude that fits the facts. For this purpose, methods from statistical physics – which were built to cope with the random behavior of atoms – are very useful.

The insight that the economy allows workers to be specialists goes back to Adam Smith's 1776 book *The Wealth of Nations*.[18] The theory of ecology, which was developed in the middle of the twentieth century, is essentially a theory about specialists. Grass is a specialist at the job of converting sunlight, soil and water into grass. Zebras are specialists at the job of eating grass and turning it into zebras, and lions are specialists in eating zebras and turning them into lions. *Homo economicus*, who can solve any problem, does not need to specialize. In contrast, the boundedly rational agents of complexity economics are specialists who use their limited abilities for what they do best. In Chapter 3 we will develop the analogy between ecosystems and economies and show how it can be used to explain an important aspect of economic growth. And in Chapters 10 and 11 we will show how it can be used to explain market malfunction, providing tools for regulators to anticipate and prevent (or at least ameliorate) financial crises.

Many of the ideas for complexity economics were articulated in the 1950s by Herbert Simon, a remarkable polymath and a key innovator in the fields of economics, artificial intelligence, cognitive psychology, management, public administration and political science. Even though Simon won the 1978 Nobel Prize in economics, the mainstream never followed his lead, and complexity economics has so far remained at the periphery of the economics profession. You might ask, 'If complexity economics has been around so long, why didn't it already take over the mainstream?' The short answer is that in Simons's day we lacked the computer power, data and understanding of human decision-making necessary to execute his vision. Fortunately for us, computers are now roughly a billion times faster than they were then, and our data capabilities are similarly greater. Just as computer scientists have used this new power to make great leaps forward in artificial intelligence and machine learning, complexity economists are beginning to take advantage of it as well. In the past five years or so this has allowed complexity economics to transition from a qualitative to a quantitative science, and we are in the process of moving from the conceptual to the practical phase of model deployment.

While we complexity economists sit on the fringe of academic economics, our models are beginning to be used for commercial purposes. Some of our ideas are beginning to inform models used by central banks to monitor financial markets in order to avert the next meltdown. In other applications under development, our models could help deal with rising inequality or find the best paths for developing countries. They can predict responses to such major transitions as automation or the switch to green energy and help us make those transitions more smoothly.

A concrete example where complexity economics can make a big difference is climate change, which is perhaps the most urgent problem the world faces right now. The source of the problem is the economy, which produces and uses the fossil fuels that create greenhouse gases. We have good models for why the global climate is warming – models that are convincing to almost all scientists. In contrast, economic models for preventing climate change are highly controversial, yielding remarkably divergent answers. The 2018 Nobelist William Nordhaus, using a standard macroeconomic model, has argued that the economic cost of climate-change mitigation is so high that we should make changes relatively slowly and allow the Earth to eventually warm by 3.4°C (about 6°F).[19] In contrast, my colleagues and I have shown that rapidly converting to renewable energy within a couple of decades will help keep global warming closer to 1.5°C, while likely saving us many trillions of dollars (see Chapter 13). Resolving this controversy, and doing so quickly, is essential to our future!

The new science of complexity economics involves a radical change in how economics is done. The fact that the intellectual framework is so different means a complete makeover of tools, skills and knowledge base – a cultural sea change. Not surprisingly, such a momentous shift is strongly opposed by the academic mainstream, as revolutions always are. For example, even though it is both innovative and useful, the model for the COVID pandemic described in the prologue could never be published in a 'top' economics journal. Why? Because it cannot be solved mathematically and, most importantly, it lacks utility maximizing agents, and thus violates the central dogma of standard economic theory. The academic departments where complexity economists work are typically *not* economics departments, but rather have names like Computational Social Science or, in my case, Geography and the Environment.[20]

Despite these barriers, complexity economics is beginning to get support from institutions that are unhappy with the conventional tools, including a few avant-garde central banks. Commercial applications are under development. With modern technology we can now accomplish things that early pioneers like Herb Simon could only dream of. The time is ripe for complexity economics to take off.

Bringing complexity economics to fruition will require substantial time and resources. Many challenging problems still have to be solved. Nonetheless, the chances of success are high and the payoff could be enormous. We need to make this revolution happen as soon as possible; the fate of the world hangs in the balance.

PART I:

What is Complexity Economics?

I should admit that I have very little formal education in economics. I was a physics major as an undergraduate at Stanford. My only encounter with economics was in 1971, during spring of my sophomore year, when I took Economics 1, a large lecture class with hundreds of students. The professor, John Gurley, was a prominent economist who was an expert on the Chinese and Soviet economies. His lectures were clear, he was a good teacher and the math was easy, but I found many of the ideas that he stated as facts hard to believe. For example, he drew two curves on the board, one for price vs supply and another for price vs demand, and stated that the price was the point where they intersected.

Because I was borrowing my way through college, I wanted to get my money's worth, and I always sat in the first row and asked questions.[1] I raised my hand and asked him how the price arrived there. If the demand curve suddenly shifted, how would the price adjust? How quickly could it move? Didn't we need to worry that it might under or overshoot? If things changed really quickly, maybe the price couldn't track at all? Professor Gurley couldn't provide convincing answers.

When I looked into this question thirty years later, I discovered that several prominent economists had worked hard to answer the same questions, without much success.[2] Economists generally simply *assume* that prices are at equilibrium, defined as the point where supply equals demand. The idea of equilibrium originated in the latter half of the nineteenth century with the work of the French economist Leon Walras. He justified how the price could get to the intersection of supply and demand via a procedure called *tatonnement* (French for 'trial and error') that was used in the Paris stock market at the time.[3] Because this process is cumbersome and time consuming it is no longer used in modern stock markets. In fact, in most modern markets transactions are made at prices where supply does not equal demand, and in some markets – like housing markets – they can be very different. As we will see, one of the contributions of complexity economics (my own work included) is in understanding how prices move when markets operate out of equilibrium, and demonstrating that this can be important.

My next encounter with economics came in the summer of 1987, at a conference called 'The Economy as an Evolving Complex System' at the Santa Fe Institute (SFI), an interdisciplinary home for the study of complexity, where I was an external professor.[4] SFI has grown since then, but it is still a small place, with about eight residential faculty members, though it has a much larger external faculty and a steady stream of visitors. Talks are given almost every day, on topics ranging across all of science, from anthropology to physics to zoology. At lunchtime or tea there are typically people from many different disciplines, and the discussions are often wide ranging, stimulating and fun.

This particular meeting was organized by two Nobel Prize winners – Stanford economist Kenneth Arrow and condensed-matter physicist Philip Anderson – who had invited ten economists and ten scientists from other disciplines (physics, biology and computer science) to participate. We met for two weeks in the former Christo Rey convent, which the Institute had just rented – its first physical location after two years as a virtual institution. The meeting took place in the chapel, which retained its stained-glass windows, but with the altar and cross removed and replaced by the speakers' lectern. At the time, I was a staff scientist at the Center for Nonlinear Studies at Los Alamos, and economics was not a particular concern. However, I was curious, and I was honored by the invitation, so I accepted.

In the first week, the economists gave us a fascinating crash course in economics. My impression was that they assumed that, since physicists are good at mathematics, we would team up and help them solve their math problems. This was not what happened. We physicists were fine with their mathematics and didn't have much to add, but we were skeptical of their assumptions. At that point in time, the theory of rational expectations still dominated economics. Then, as now, all standard economic theories assumed some sort of equilibrium. The physicists had a reflexive reaction against both notions. To us, economic equilibrium was a potentially dangerous approximation and rational expectations seemed completely unrealistic. Our view was that people are not consistently rational and that economic theory needed to take human psychology into account. The nascent field of behavioral economics was attempting to do just that, but at that time it was still off the radar of the economists at the conference. There was consequently a great deal of arguing. Physicists tend to view themselves as the kings of the physical sciences, and economists view themselves as the kings of the social sciences, and they can both be arrogant. (At one point, Phil Anderson, who was famous for being shockingly blunt, exclaimed to the economists, 'You don't *really* believe that, do you?!' and Harvard economist Lawrence Summers weighed in with 'You physicists all have a Tarzan complex.') Despite the tensions at the meeting, I learned an enormous amount and came away with a renewed interest in economics. It was obviously a field full of contradictions and bad assumptions, suggesting that it could be a rewarding place to pursue innovative research.

My interest in economics grew even more after meeting the Yale economist John Geanakoplos, whose friends all call him John G., when he directed the economics program at SFI in 1990–1991. In those days, there was a weekly SFI basketball game at a nearby park, and John G. and I were both regulars, which helped us get to know each other. One of our favorite pursuits was to argue about the pros and cons of the founding principles of economics. John taught me an enormous amount, and our dialogue eventually led to a 2009 paper called 'The Virtues and Vices of Equilibrium and the Future of Financial Economics', which presents a synthesis of our converging view that 'economics needs to move in new directions if it is to continue to make progress'.[5]

1.

What is a Complex System?

Complex consequences can arise from simple causes, especially when you allow the system you are looking at to be at a different scale from the elementary entities from which it is made.

Phil Anderson (2000)[1]

The nature of complexity

Complexity economics is the application of ideas from complex systems to economics. But what is a complex system?[2]

In my physics courses at Stanford, I learned a great deal about motion, force and energy. My courses took a reductionist approach: The goal was to understand the fundamental rules that determine the interactions of simple things – like masses on springs or elementary particles – and study them in simple contexts. In celestial mechanics, for example, we learned how to calculate the orbits of two bodies, such as a single planet and the Sun, but more complicated situations, such as the orbits of three such bodies, were omitted, because the equations couldn't be solved.

I couldn't help but ponder the big questions that my courses were not addressing. I wanted something that was the opposite of reductionism. Given that all the fundamental forces of nature are simple interactions between relatively simple physical things, how can this lead to phenomena such as life and intelligence? How did matter spontaneously organize itself in the origin of life? How did evolution create such a remarkable thing as the human brain?

When I was in graduate school at UC Santa Cruz, I discovered cybernetics, which was the precursor of complex-systems theory. My friends and I read and discussed Norbert Wiener's ideas about feedback and control, Claude Shannon's theory of information, John von Neumann's self-reproducing automaton and Erwin Schrödinger's reflections about life and the mind.[3] The big question we pondered more than all others was the mystery of self-organization: How can disorganized configurations of matter spontaneously become organized?

Later, at Los Alamos, I organized conferences and workshops on 'Cellular Automata', 'Evolution, Games and Learning' and similar topics, and invited leading scholars who were thinking about these things to visit, creating a hub of activity.[4] The senior fellows at Los Alamos decided that this kind of work needed its own institution, free from the taint of a weapons laboratory. So, in 1984, the Los Alamos physical chemist George Cowan co-founded (with theoretical physicist Murray Gell-Mann) the Santa Fe Institute, and shortly afterward the phrase *complex systems* entered our vocabulary. I became affiliated with the Institute, and, in 1988, I started the Complex Systems Group at Los Alamos and recruited several outstanding postdoctoral researchers who later became leaders in the field. There are now thousands of scientists devoted to the study of complex systems, scattered across almost every discipline in the physical, natural and social sciences, including economics.

What is a complex system? By definition, a system is complex if it has emergent properties.[5] I've already mentioned the example of the brain. We humans don't just think – we are conscious, with a sophisticated model of ourselves, the world around us and how we fit into it. Consciousness isn't present in the individual building blocks – it emerges from the interaction of billions of neurons. In human brains and other complex systems, emergence happens when building blocks are connected together to give rise to behavior qualitatively different from what any of the building blocks can do alone. Ant colonies are another classic example.[6] Individual ants are simple creatures with only about 250,000 neurons, so the reasoning power of individual ants is extremely limited. But ants communicate through chemical signals and touch. This gives a colony of ants computational powers that individual ants lack, and through their collective interactions with one another they can do surprisingly sophisticated things. Ant colonies cultivate fungi, raise aphids, fight wars and take slaves. Individual ants can't do these things. There is no Super Ant directing the colony – the queen is just an egg factory. Instead, these sophisticated collective behaviors emerge from the interactions of many ants.

Emergent behavior can be counterintuitive. The idea that thought and consciousness are purely mechanical processes, that can be explained from the bottom up, is so radical and difficult to fathom that it was not widely accepted until the twentieth century. Given that farming was not invented by *Homo sapiens* until what we term the agricultural revolution of about 10,000 years ago, how could simple creatures like ants have been doing something similar for millions of years?

Complex systems like the brain or an ant colony are called *adaptive complex systems*. They are distinguished from ordinary complex systems with simpler emergent behaviors by the fact that their properties have evolved over time, through a process of selection. Consciousness doesn't emerge by simply hooking up neurons at random; they have to be connected in just the right way. It took evolution billions of years of trial and error to invent the human brain. Thinking in evolutionary terms is essential for understanding adaptive complex systems. And evolution is not just about biology; it applies to a broad range of adaptive complex systems, including economics and the other social sciences. Firms, for example, have to make profits or they quickly go out of business, illustrating that selection underlies the essence of how capitalist economies function. The study of adaptive complex systems has helped clarify the basic principles of evolution so that they can be applied more generally outside of biology.

Can we make better economic predictions?

Through science, things that appear to be random can be rendered predictable. Prediction is enormously useful because it allows us to make sensible decisions. We tend to take for granted the myriad ways that science has fundamentally changed our view of the world. The Trojan War from Greek mythology provides an example: After Agamemnon has gathered his troops to fight Troy, the wind refuses to blow, so his fleet cannot set sail. He consults Calchas, the seer, who tells him that he must sacrifice his beloved daughter Iphigenia to appease the goddess Artemis. He has his daughter killed, and the winds begin to blow.

Modern science casts a different light on this story. Unlike the ancient Greeks, we see no conceivable causal connection between Iphigenia's sacrifice and whether or not the wind blows. Today, a father who acted as Agamemnon did would be regarded as utterly mad. A modern Agamemnon would simply consult the weather forecast, note the high-pressure zone hovering over Argos and tell his men to wait a few days until the wind started to blow again. Physics has transformed something formerly believed to depend on the whims of the gods into a predictable event whose causal mechanisms are well understood.

Fast forward to contemporary Greece, where an example shows that we still lack broad agreement about essential cause-and-effect relationships in economics. Since 2008, the Greeks have suffered one of the worst depressions in modern times. Greek GDP dipped by 30 per cent from its peak, unemployment reached 27 per cent, and in 2021 almost half of Greek young people (aged 15–24) were still out of work. Paul Krugman and many others have argued that all this suffering could have been avoided if only Germany had not forced Greece to adopt a policy of austerity (budgetary belt-tightening) just when economic stimulus was most needed.[7] In contrast, Wolfgang Schäuble, the German finance minister, along with a host of prominent economists, insisted that austerity was unavoidable and the only way for Greece to become solvent again.

Who was right? There is no widely agreed-upon economic model that can answer that question. This is not unusual: While the majority of economists agree on some things, in many cases an array of different models make divergent predictions, leaving people free to choose the model they like best, based on ideology, politics and self-interest.[8] We can break out of this stalemate only by finding models whose predictions are consistently good enough to gain a wide level of credibility.

My goal in this book is to set out a vision for how we can build models that make better economic predictions. Surprisingly, many economists do not agree on the importance of this goal. To quote two prominent Harvard economists, David Laibson and Xavier Gabaix,

> Predictive precision is infrequently emphasized in economics research [. . .] In this sense, economic research differs from research in the natural sciences, particularly physics. We hope that economists will close this gap. Models that make weak predictions (or no predictions) are limited in their ability to advance economic understanding of the world.[9]

When I emphasize this in my talks, economists often respond that the central goal of economics is to provide a conceptual framework for thinking about the world and evaluating policy choices. I wholeheartedly agree that having a conceptual framework is central, but, in my view, this only makes the quest for better predictions even more important. If we can't make reliable predictions, then how do we know if the conceptual framework is correct?

It is easy to forget that prediction is interwoven into everything we do. To pick up a glass of water, your brain has to predict how your muscles need to be stimulated to get your hand to move to the glass and grasp it, then predict how to compensate for the weight of the glass as you bring it to your lips to drink. Similarly, every policy decision represents a prediction that the economy is likely to do better if that choice is made.

I should be clear about what I mean by the word *prediction*. Niels Bohr is said to have declared that 'prediction is very difficult, especially if it's about the future'. This remark seems ironic; aren't all predictions about the future? Actually, no, not at all. Consider Boyle's Law: In 1662, Robert Boyle invented a device that allowed him to control the volume of air inside a container and showed that the air pressure is inversely proportional to the volume. In other words, if you know the volume, you can predict the pressure, and vice versa. This is scientific prediction, true at any point in time – present, past or future.

Of course, many predictions *are* explicitly about the future. For instance, central banks and treasury departments use models to make predictions about GDP and unemployment. Models provide a sandbox where they can test policy options to understand how well they work and how they might be revised to work better. What happens if we raise interest rates? What happens if we require banks to hold more capital? What happens if we raise the minimum wage? Implement a universal basic income? Decrease taxes? Leave everything as it is? Bad models lead to bad decisions.

Predictions are also important because they test models and tell us whether we can trust them. Does the model we're using capture cause-and-effect relationships well enough? If the model fails to make accurate predictions about the world as it is now, we can't trust it to correctly answer the 'what-if' questions we need to pose when contemplating a change in policy. Unfortunately, even after decades of hard work by several generations of macroeconomists, Nobel Laureates in economics like Paul Romer question whether the effectiveness of macroeconomic forecasting has improved at all.[10]

Can we do better? Many argue that this is impossible for several reasons: As I've already explained, economics is about people, and, because people can think, their behavior is hard to predict. Or there may be fundamental limits: The economy may have intrinsic properties, such as market efficiency or chaos, that defy accurate prediction by any model, no matter how good.

We will never be able to predict the economy perfectly – far from it – but we can do much better than we are doing now. To return to Niels Bohr: Yes, prediction *is* difficult. But there are plenty of examples of good predicting, even of the future. Here's one from my own experience, which has contributed to my belief that, if we change our approach, we can make much better predictions in economics.

The Project: Using physics to beat the house

My journey to complexity economics began more than four decades ago in an unlikely way – by using physics to beat the game of roulette.[11] How this came to happen is a good example of 'sensitive dependence on initial conditions', a concept we will explore in more detail later. I grew up in Silver City, New Mexico, which was once the home of the infamous gunslinger Billy the Kid. When I was twelve years old, due to a complicated chain of circumstances involving Joe McCarthy, Frank Oppenheimer and a mythical gold mine in west Texas, an idiosyncratic young genius named Tom Ingerson moved to Silver City and became the sole member of the physics faculty at Western New Mexico University.[12] Tom also started Explorer Post 114 – a branch of the Boy Scouts for teenagers – which I joined, along with my friend Norman Packard, who was the other science kid in Silver City. In addition to inspiring Norman and me to become physicists, Tom had entrepreneurial aspirations that infected us as well. Norman's name will appear many times in this book – he is a remarkably creative person, and I have had the good fortune to have him as my partner in several endeavors.

Fast forward to the summer of 1975, when I was starting my third year of graduate school at the University of California at Santa Cruz, specializing in cosmology and doing research on galaxy formation. I spent the summer working for the Forest Service in Libby, Montana, where poker had just been legalized. I read *The Complete Guide to Winning at Poker*, began playing in the evenings and soon discovered that I was making more money gambling than I was in my day job. Meanwhile, Norman was spending the summer playing blackjack in Las Vegas; in our letters we debated whether poker or blackjack was the best way to supplement our incomes. When I met Norman at the end of the summer at the Greyhound bus station in Portland, Oregon, where he was finishing his undergraduate degree at Reed College, he convinced me that we could make our fortune using physics to beat roulette.

'How is that possible?' you might ask. A roulette wheel is designed to be a random-number generator. The croupier sets the central rotor spinning in one direction and the ball in the other. Initially the ball spins so fast that it runs along the outside of the beveled track, but eventually it slows and spirals into the center and bounces around on the numbered cups until it comes to rest in one. Because the relationship between the motion of the wheel and the motion of the ball seems unpredictable, the final number also seems unpredictable, making the sequence of numbers generated by a roulette wheel effectively random.[13]

Or is it? From the point of view of a physicist, roulette is a simple exercise in classical mechanics. According to Newton's laws, the forces acting on an object, together with a measurement of its position and velocity, allow us to predict the object's future motion. The main force acting on the ball is air resistance.[14] About ten or fifteen seconds elapse between when the croupier releases the ball and when she closes the bets. This provides an opportunity to measure the position and velocity of the ball in one direction and the wheel in the other, use physics to make a prediction about where the ball will land and place bets.[15] Doing this gave rise to a five-year undertaking we called The Project, which over its history enlisted the efforts of about thirty people. We organized a front company, Eudaemonic Enterprises, to buy sophisticated electronic parts. (Eudaemonia is an Aristotelian notion, which our ancient Webster's Dictionary defined as 'a state of felicity or bliss obtained through a life lived in accordance with reason'.) Our dream was that the fortune we would win at roulette would allow us to be independent of the system, so that we could start a eudaemonic science commune and do research on what has now come to be called complex-systems science.

Norman joined me at UC Santa Cruz to do his graduate work in physics. We bought a roulette wheel and installed it in the living room of our communal home, a sprawling California-style bungalow, where our work was punctuated by shared meals, intense intellectual discussions and wild parties. We solved the equations of motion, made thousands of measurements and confirmed that it was possible to predict roughly where the ball would land on the wheel. The subsequent bouncing motion was too complicated to predict, but after careful study we determined that when the ball leaves the track, it is unlikely to come to rest on the opposite side of the wheel. Our predictions were far from perfect, but they were still much better than random, giving us a substantial edge over the house.

But how would we cash in on these predictions in a casino? We needed to make our calculations in secret and quickly enough to place bets, but in 1976 there was no such thing as a personal computer, much less a smartphone. Following some insightful suggestions from our mentor Tom Ingerson, we designed and built the world's first wearable digital computer. It was based on the 6502 microprocessor, the same one used to build the original Apple computer.[16] With the technology available at the time, there was barely enough space to fit 3,000 bytes of program memory into the biggest box we could hide. I painstakingly hand-coded the program in machine language; this included an operating system (with perhaps the shortest floating-point arithmetic package ever written) and all the logic needed to perform the calculations and output the results. Our first wearable computer was named Harry, the middle name of its builder, Norman H. Packard. Finished in 1977, it was concealed under the armpit. The other armpit concealed a pack of a dozen AA batteries, which allowed the device to run for about an hour and a half. The user wore a complex body-harness carrying the computer, battery pack, wires, switches and antenna, along with vibrators that tapped out the outputs. Our second version was called The Sandwich because it had two PC boards, like two pieces of bread, glued together with a filling of microcrystalline wax. It was finished three years later and built into a shoe, which held the batteries, switches, antenna and vibrators. (Only users like me with large feet could operate it.)

At the roulette table, one teammate, usually me, measured the velocity of the ball and wheel by clicking switches embedded in his shoes and operated by his big toes. I drew on some aspects of my brief experience as a professional poker player, using the pseudonym Clem, acting stupid and betting small stakes. My computer calculated its prediction and transmitted a signal to an accomplice wearing a second computer – this one named Renata, in honor of its builder, Ingrid R. Hoermann. The computer and its vibrator were built into the armpit area of her bra (which provided a convenient strap to mount things on, and which we thought had security advantages); she then placed bets on numbers near where the ball was most likely to land. We consistently beat the house, achieving a 20 per cent edge, but due to chronic hardware failures and fear of broken kneecaps, we never managed to play for really high stakes and didn't get very rich as a result.

However, I learned many valuable lessons. The Project taught me the scientific method, drilling it deep into my soul in a way that graduate school never could. We developed the theory, designed the experiment, executed some difficult engineering and tested our hypotheses both in the lab and in the casino. It also taught me that randomness is a subjective concept that depends on one's state of understanding. Better science and better technology can make something that seems random become predictable.

The economy appears remarkably unpredictable, but might this just reflect our current state of misunderstanding? Economic variables – like GDP, unemployment and inflation – change all the time, in seemingly random ways. But maybe the economy is like a roulette wheel – perhaps its changes only seem random because we aren't thinking about their causes in the right way. If we built better models, using new ideas and making better use of computers and big data, could we gain a new understanding that would allow us to make better predictions?

Could chaos explain business cycles?

The major changes in the economy are called business cycles. During expansionary periods, GDP grows quickly and unemployment is low; during recessions, GDP falls and unemployment is high. These oscillations are irregular both in their timing and their intensity; in the United States, the average span of a cycle is about seven to ten years, though there is a great deal of variation. A central goal of macroeconomics is to understand what causes business cycles and how to avoid or at least reduce the intensity of recessions.

Remarkably, the underlying causes of business cycles remain controversial. As noted in the Introduction, changes in the economy fall into two basic categories: those that come from outside the economic system, and those that are generated within. The explanation offered by standard economic models is that changes in the economy are initiated by outside influences, which economists call *shocks*. To me, this word evokes the image of Zeus sitting on Mount Olympus, hurling lightning bolts at the Earth and disturbing the mortals below. Zeus is unpredictable, and the shocks arrive at random. The COVID pandemic provides an uncontroversial example where a huge economic change was caused by an outside shock.

The alternative explanation is that change comes from within, or, in more technical jargon, motion is endogenous: The economy changes on its own due to internal factors. In reality, it is a mixture of the two – sometimes the economy responds to outside factors, sometimes changes arise from within, and sometimes both types of causes act at the same time. Standard models, however, generally attribute *all* economic change to outside influences. According to these models, absent outside shocks, the economy would settle into a state of rest. It changes only because Zeus keeps shocking it, knocking it away from equilibrium. Shocks are random events from outside the economy that are by definition unpredictable. All that standard economic models aspire to do is to predict how the economy will respond to the shocks *once they happen*. This places a strong limit on their usefulness.

Yet there is a perfectly plausible explanation for how the economy can change from within. This is the beautiful mathematical concept now known as *chaos*, which is an important component of complex-systems theory. Economists in the 1980s and 1990s briefly experimented with the idea that chaos caused endogenous economic change and then they dropped it. To my mind, it was dropped for the wrong reasons, and it's time to bring it back.

What is chaos?

One evening in 1977, Robert Shaw, a fellow physics graduate student at Santa Cruz, appeared at our usual communal dinner in a state of excitement. He had just been introduced to meteorologist Edward Lorenz's mathematical model, which featured a system of three simple equations and contained something called a *strange attractor*.[17] Rob had programmed the equations on the computer – did we want to take a look? Rob tried to explain why he was so excited to Norman's girlfriend, Lorna, who was not a physicist. The explanation went like this:

Models that describe how things change over time are known as *dynamical systems*. The basic idea goes back to Newton, who showed that you can predict the movement of any physical system if you know the state of the system and the forces acting on it. In physics, the state of the system is the positions and velocities of its components, which could be planets or roulette balls. We say the dynamical system is *deterministic* if no randomness is involved; that is, if the state at a given instant in time precisely determines the state at the next instant. As the state evolves, it traces out a curve, called the *trajectory*.

Most dynamical systems have what are called *attractors* that determine their long-term behavior. For example, due to air resistance, a pendulum will eventually come to a state of rest. This type of attractor is called a *fixed point*, for obvious reasons. For a slightly richer example, consider a grandfather clock, whose pendulum has a spring that keeps it moving. As long as the spring is wound, the pendulum swings back and forth at a consistent frequency; the attractor here is a pattern of oscillation, called a *limit cycle*, which looks like a closed loop: as the pendulum swings back and forth, a plot of its position vs velocity traces out a circle, which it goes around again and again.

As a physics graduate student, I had encountered quite a few deterministic dynamical systems in my textbooks. My view at the time was that the words *deterministic* and *predictable* were almost synonyms. But my experience with roulette had caused some doubts to tickle the back of my brain: How could a system like a roulette wheel – which, as a physical system, is by definition deterministic – also be a random-number generator? What was it about roulette that made it so much harder to predict than planetary motion?

As soon as dinner was over, we got in Rob's 1957 Ford station wagon and drove to the university, so that we could use the physics department's analog computer. Rob had programmed the Lorenz equations, which are a highly simplified model of atmospheric convection. They form a simple dynamical system with three equations relating three variables – three numbers labeled x, y and z representing the current state of the atmosphere. Its state can be visualized as a point with coordinates (x, y, z) in a three-dimensional space. We could see the action of the dynamical system by projecting this point on an oscilloscope screen and watching it move. Rob picked some initial values for x, y and z, pushed the Start button, and we watched the trajectory of the Lorenz equations unfold as x, y and z changed over time. What we saw was completely different from anything we had ever encountered in our textbooks. The trajectory traced out a beautiful pattern, like a butterfly whose wings fold back into each other, as shown in Figure 1.

The motion never stopped, and the pattern *never repeated itself exactly*. The pattern was an attractor – Rob made this clear by starting from several different initial conditions and showing that the trajectory always settled into the same shape – but the nature of this attractor was very different from the fixed points and limit cycles we were used to. It was appropriately named a *strange attractor*.

An attractor of this type is now called a *chaotic attractor*, and the motion it produces has come to be called *chaos*.[18] Chaos is characterized by two essential properties: The first is *sensitive dependence*

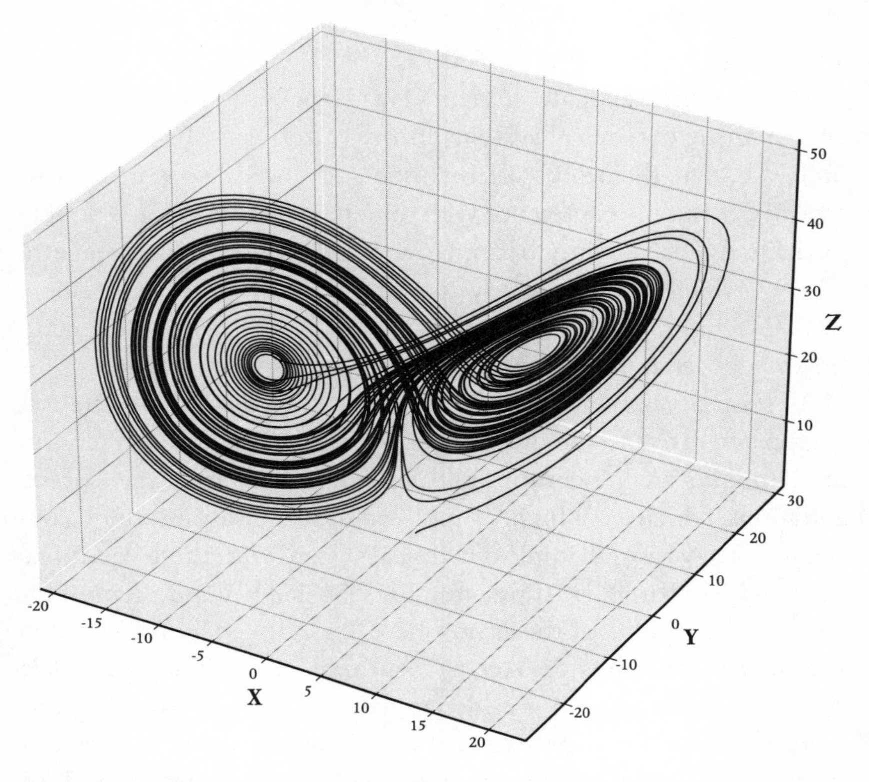

Figure 1. *The Lorenz attractor.* What is shown is a single trajectory.

on initial conditions, and the second is *endogenous motion*, meaning that, even though there are no external shocks, the system never settles down to rest. That said, the word *chaos* can be misleading, because chaos in the mathematical sense does not imply chaos in the same sense that we use the word in everyday English. My PhD thesis was titled 'Order Within Chaos', and there are many circumstances in which order and chaos coexist. In fact, this is typical and will become important later when I argue that if we understand the chaotic properties of the economy, we can make better predictions.

The concept of chaos was anticipated by the brilliant French mathematician Henri Poincaré at the end of the nineteenth century. His inspiration came from studies of the three-body problem in celestial mechanics (for example, the motion of the Earth, the Moon and the Sun). He used sensitive dependence on initial conditions as a metaphor to explain how 'fortuitous phenomena' can have such a huge effect on our lives.[19] But he was working in the days before computers, which made chaos almost impossible to study, so he couldn't get very far. The concept of chaos was virtually forgotten until computers were invented and Lorenz rediscovered it.

To demonstrate sensitive dependence on initial conditions, Rob did a series of experiments. As before, he picked initial values for the state (x, y, z), hit Start to trace out a trajectory, then stopped the computer, freezing the pattern on the screen. Then he set the computer to approximately the original starting values of (x, y, z) and repeated the experiment. At first, the trajectory retraced itself, apparently exactly as before, but after a while it began to drift away and then diverged completely from the original trajectory. Remarkably, the experiment was not reproducible; that is, even though the initial values were almost the same, the outcome was very different – the final state ended up in a completely different place. This was really surprising, because the Lorenz equations are *deterministic*; the initial conditions should *exactly* determine all future values. So why wasn't the experiment repeatable?

The difference arises because of the combination of uncertainty in the initial conditions and the unusual geometry of the Lorenz equations (and other systems with strange attractors). It is never possible to specify initial states with *perfect* precision. Every time Rob reset the computer, the initial state was a little bit different, even if this difference was imperceptible to us. For regular motion, where the attractor is a fixed point or a limit cycle, this difference doesn't really matter: Even after a long time, the initial differences in the states remain imperceptible and the simulation is repeatable far into the future. But when the dynamics are chaotic, the geometry of the dynamical system stretches and folds the trajectories, amplifying these small differences exponentially, so that the final outcome eventually differs dramatically.

Rob also gave us a visceral illustration of the difference between regular motion and chaos. The computer output was put into a speaker, so that we could listen to the motion of the Lorenz equations in addition to watching it. For a fixed-point attractor, the computer was silent. For a limit-cycle attractor, the sound was a pure tone, like someone whistling a single note. But for a chaotic attractor, the sound was white noise, like the roar of a windstorm, suggesting its connection to fluid turbulence, which I will explain in a moment.

I had been wrestling with the difficulty of predicting roulette for more than a year, and seeing the Lorenz attractor caused an explosion to go off in my head. Because the ball eventually comes to rest, the motion of the ball has a fixed-point attractor. Nonetheless, on its way to its resting point, it displays sensitive dependence on initial conditions, which is why the final resting point is so hard to predict. While the ball is rolling on the track, the motion is reasonably predictable, but as soon as the ball falls onto the wheel, small differences in initial conditions can make a huge difference in its final resting point. When the ball hits the wheel, it might strike a cup head on and bounce backward, but if it hits just a millimeter farther, it might bounce forward and land in a different place. Tiny changes are greatly amplified. The roulette wheel is a good example of sensitive dependence on initial conditions. The sensitivity wasn't strong enough to take away our advantage, but it didn't make our job easy.

After Rob showed us the Lorenz equations, Norman and I were eager to get to work researching chaos as soon as possible. After eleven trips to Nevada, building and fixing roulette computers and taking heat in the casinos, we were burned out. While I loved cosmology, the problems to be solved were well known and had all been worked on by very smart people. In contrast, back then, many basic aspects of chaos were not yet understood; it was a wilderness waiting to be explored. I had always been interested in what causes order as opposed to disorder. Chaos is a mechanism for creating disorder in an orderly way, and it gave me the entrée I had been searching for.

Norman and I teamed up with Rob and a brilliant undergraduate named Jim Crutchfield and formed our own research group, which we called the UCSC Dynamical Systems Collective. Strange attractors were far from what other physicists were thinking about at the time. Although a few supportive faculty members thought the idea was interesting, no one knew enough to supervise our PhD work. We went ahead nonetheless, sharing authorship on all our papers and co-supervising each other.

We set out to prove Lorenz's hypothesis that turbulence is caused by chaos. Most of us have experienced turbulence directly on plane flights – or in hurricanes, which are an example of fully developed turbulence. Or perhaps you have sat by a mountain stream, watching the water as it burbles and gyrates and spills over the fixed rocks in the stream-bed, moving by itself, as if the water had a will of its own. Changes in weather are also caused by turbulence; even though the continents and oceans are heated and cooled in a regular seasonal manner, the weather is irregular and difficult to predict.

In 1978, when we began our research, the prevailing theory was that turbulence happened because 'things are really complicated'. Lorenz proposed the alternate theory that turbulence is caused by chaos. Though this is now well-accepted, it was far from obvious at the time. While his equations were derived to explain fluid flow (the motion of gases, like air, as well as liquids), they were based on extreme approximations. Most researchers studying turbulence at the time had never heard of chaos, and when they did, they were skeptical.

There is a broader lesson here. Although complex systems *can* be complicated, they don't *need* to be complicated. The words *complicated* and *complex* are not synonyms! *Complicated* means that something has many moving parts, whereas *complex* means it exhibits emergent behavior – that is, it does things that are not easily predicted *a priori*. The Lorenz equations provide an excellent example of how a very simple set of equations, with only a few terms, can give rise to a complicated geometric object (the attractor) with complicated motion. The attractor is complicated but its underlying cause is simple. Even though the equations are simple, there is no formula that gives the answer – it is only possible to solve them by simulating them on a computer.

We of the Dynamical Systems Collective set out to invent new mathematical methods that could be used to identify and characterize chaos in the real world. In 1980, we wrote a paper titled 'Geometry from a Time Series', with Norman as the lead author, which took the first step toward accomplishing this.[20] Experiments are rarely able to observe all the variables needed to fully determine the state of their underlying dynamical system. We found a way to reconstruct the missing information. For example, in fluid-flow experiments the data often consisted of the velocity of the fluid in a given direction at a given point in space. We showed how to use this kind of limited data to calculate key statistics about chaos and draw pictures of the underlying chaotic attractors. A couple of years after the publication of our paper, our methods were used to demonstrate the existence of chaos in chemical reactions and weakly turbulent fluid flows, and since then they have been employed thousands of times to show that chaos exists in many different phenomena, ranging from dripping faucets to heart arrhythmias to imperfect electronic control systems.

Is the economy chaotic?

> *If you hit a rocking horse with a stick, the movement of the horse will be very different from the stick. The hits are the cause of the movement, but the system's own equilibrium laws condition the form of movement.*
>
> *Knut Wicksell (1918)* [21]

A problem with prevailing macroeconomic models is that they have only fixed-point attractors, so if the model economy is left alone it settles into a state of rest and stays there forever. This implies that all changes in the economy are driven by events outside the economy. To explain why the economy changes, economists have to postulate the existence of shocks that knock the model away from equilibrium. They cannot predict the shocks, they can only predict how the economy relaxes back to equilibrium. The Swedish economist Knut Wicksell, quoted above, uses the analogy of a rocking horse being struck by a stick: The rocking horse is like the economy and the blows from the stick are like the shocks. Absent further shocks, the economy would settle into a static equilibrium, but the blows from the stick keep coming, and the horse continues to rock as a result.

There have long been dissenters who felt that a substantial component of the economy's changes come from within. In the 1960s and 1970s, several prominent economists proposed theories for business cycles based on deterministic models with limit cycles.[22] In these models, the economy is like a rocking horse ridden by a child, who pumps it back and forth and keeps it going indefinitely – like the spring powering the pendulum in a grandfather clock. The problem with limit cycle-based models is that their business cycles are too regular. These models produce limit cycles that are perfectly periodic, like the motion of a rocking horse, whereas real business cycles are highly irregular; the intervals between cycles can range from several decades to only a few years. Models with limit cycles can be made less regular by adding shocks, but it's still hard to produce behavior irregular enough to be plausible (though this point has recently been contested).[23]

As the idea of chaos became better known, a few economists asked whether the irregular oscillations of the economy might be due to chaotic dynamics. If it's true for the weather, they reasoned, why not the economy? Unlike limit cycle–based explanations, chaos automatically generates irregular behavior, and it's easy to find examples of simple chaotic dynamical systems whose output resembles business cycles. In 1981 Richard Day proposed a modification of an economic-growth model with chaotic dynamics, and in 1986 Michele Boldrin and Luigi Montrucchio gave another example.[24]

These models were highly simplified and required unrealistic parameters to generate chaos, but they at least suggested that the idea was plausible. But did chaotic dynamics figure in *real* economies?

This question led the economists William 'Buz' Brock, Blake LeBaron and José Scheinkman to formulate a test for identifying chaos in economic data, building on methods we had developed in the Dynamical Systems Collective. They applied their test to a long historical record of daily changes in the Dow Jones Index and showed that there were no convincing signs of it.[25] A few other papers examined different economic data and got similar results. From the point of view of mainstream economists, this closed the case: Since there was no empirical evidence for chaotic dynamics in the data, chaos must be irrelevant for economics.[26]

I like Brock, LeBaron and Scheinkman's method – in fact my collaborators and I later built on it to debunk the idea that simple chaos explains fluctuations in brainwaves.[27] However, I think the interpretation of their results was wrong. They don't rule out the possibility that chaotic dynamics underlies economic dynamics.

This misinterpretation stems from a subtle but profound point that is not widely understood: Chaos can be simple or complicated. Simple chaos and complicated chaos lie at two ends of the same spectrum, but their behavior is quite different. In simple chaos (like the Lorenz equations), the motion can be described using only a few variables. For complicated chaos, like the weather, many factors act independently, and many variables are required to describe the motion. For instance, models for predicting the weather depend on the air temperature, pressure, humidity, cloud cover, wind speed and direction, not to mention the local topography, the height and type of the clouds, the current amount of rain, the temperature of the land, and so on. Because the list of numbers needed to describe the current state of the weather is long, we say it has many *degrees of freedom*. While the Lorenz equations were meant to describe something about fluid turbulence, they do not even begin to describe all the complicated features of real weather. There is a continuum between simple models like the Lorenz equations (which have only three degrees of freedom) and the very complicated models used to predict real weather (which have billions of degrees of freedom). In my PhD thesis, I was the first to study the progression from simple chaos to complicated chaos and show how chaotic behavior changes and becomes more turbulent as the number of degrees of freedom increases.

The standard economics 'rocking horse' model is an example of a family of models widely used throughout applied science and engineering called a *random process* (in other fields, the shocks are usually called *noise*). In a random process, the noisy inputs are modified by deterministic components of the process – just as the rocking horse responds to the blows by the stick that shape the resulting behavior. Like complicated chaos, random processes have many degrees of freedom. This can make the two very difficult to tell apart.

The test that Brock, LeBaron and Scheinkman applied to the Dow Jones index was essentially a measurement of the number of degrees of freedom. If they had found few degrees of freedom, they would have been able to claim they had found chaos, which is the only way to generate random-looking behavior with low degrees of freedom. But they found high degrees of freedom, indicating that either the behavior of the Dow Jones index is a random process or it is complicated chaos – they couldn't distinguish between the two because, based on historical data alone, complicated chaos is impossible to distinguish from a random process. However, their results were not interpreted this way, but were instead taken to mean that chaos was not present, period. If they had applied their test to the weather, it would have yielded the same answer, even though we know that the weather is chaotic, for two reasons. First, we can do experiments with fluids and study the transition to turbulence. At the transition point, the motion has only a few degrees of freedom, and it becomes possible for tests like that of Brock, LeBaron and Scheinkman to show this. Second, we can simulate the equations that describe the weather on digital computers and show that chaos is present, like Lorenz did. It's far easier to demonstrate the presence of chaos in a model than it is for real data, and this is particularly true for complicated chaos.

In fact, it would have been really surprising to see simple chaos at work in economic data. The economy is inherently complicated, with many different interacting factors. Even if much of economic change is endogenous, much is not. Moreover, because the economy changes slowly, in timescales measured in years, there is effectively very little data to analyze. There have been only about ten or twelve business cycles since World War II, whereas methods for detecting even the simplest kind of chaos require thousands of cycles.

There is another strong argument for the presence of chaos in economics – one that few economists are aware of. David Ruelle and Floris Takens, who were part of a community of mathematicians in the 1970s who built on Poincaré's early insights and rediscovered Lorenz's work, proved a remarkable theorem which says that chaos is the likely explanation of complex-looking behavior.[28] The theorem is too technical to describe here, but it basically says that when dynamical systems cause endogenous motion (like limit cycles), as this motion gets complicated, it inevitably becomes chaotic (meaning that it has sensitive dependence on initial conditions).[29]

The same argument applies to the economy. If the economy produces endogenous motion that isn't a simple limit cycle or a combination of limit cycles, it must be due to chaos – there's really no other option. As noted, the models showing chaos that were proposed in the early 1980s were unrealistic to start with. In models based on more plausible behavioral assumptions – i.e., complexity-economics models – chaos becomes more common.

In 1981 I finished my PhD in physics at Santa Cruz and began looking for a job. But my specialty was now chaos. Even though our work on dynamical systems was getting a lot of attention, I didn't fit neatly into any of the academic boxes for physicists, so there were essentially no academic jobs for which I was a viable contender. Academia is, unfortunately, a world of disciplinary silos. Our research fell between the cracks – or more accurately, the chasms – that exist between disciplines. Physicists thought what we were doing was interesting, but it was utterly different from what any of them did, and there were no other disciplines doing this kind of work.

I found my job by chance: I happened to be reading a biography of J. Robert Oppenheimer when I saw a poster advertising the J. Robert Oppenheimer Fellowship at Los Alamos National Laboratory. I had grown up in New Mexico and was eager to return. Despite my misgivings about the fact that Los Alamos is, among other things, a weapons-research laboratory, I decided to apply. During the interview, they told me I didn't have to do any military research and could work on anything I wanted, and I became an Oppenheimer Fellow at the newly formed Center for Nonlinear Studies.

The intellectual environment I found myself in was remarkable. There were no disciplinary boundaries. The only questions people asked about your work were, 'Is it interesting?' and 'Is it useful?' It's no coincidence that many of the best papers in nonlinear dynamics (the broader field that chaos belongs to) were written there. The Manhattan Project had enlisted the best mathematicians and physicists in the world, and some of my heroes from that era, such as Stanislaw Ulam and Nicholas Metropolis, were still there. I had regular conversations with the mathematical physicist Mitchell Feigenbaum, a pioneer of chaos theory. Los Alamos was perhaps the only place in the United States where working on chaos was considered a perfectly normal thing to do, and with my help and that of others it became one of the places that the science of complex systems emerged from.

Chaos, which explains how randomness emerges from order, is only a small subfield within complex systems. Let's now zoom out to discuss the bigger picture.

2.

The Economy is a Complex System

In the last few years the concept of self-organizing systems – of complex systems in which randomness and chaos seem spontaneously to evolve into unexpected order – has become an increasingly influential idea that links together researchers in many fields, from artificial intelligence to chemistry, from evolution to geology. For whatever reason, however, this movement has largely passed economic theory by. It is time to see how the new ideas can be applied to that immensely complex, but indisputably self-organizing system we call the economy.

Paul Krugman (1996)[1]

Twenty-five years ago, the distinguished economist Paul Krugman joined the complexity enconomics pioneers, and proposed that the time had come to study the economy as a complex system. Unfortunately, like many before him, this largely went unheeded by the mainstream, whose conceptual models and mathematical toolkit were not well suited to take advantage of complexity theory, and Krugman himself went on to pursue other things, such as becoming a columnist for the *New York Times*.

The remarkable evolution of civilization, from the early days of *Homo sapiens* until now, is a dramatic illustration of Krugman's point that the economy is a self-organizing complex system.[2] The economy is just a name for the process of specialization, cooperation and competition that supports us.[3] The sustainable human population has grown by several orders of magnitude due to technological progress and advances in social organization, which underpin our evolving economy – without it, most of us would not exist. Because of the emergent properties of the economy, most of our lives are far more prosperous and secure than they would be if we were Robin Crusoes acting on our own. The economy's organization of our collective behavior enables us to create vaccines, predict the weather, go to the Moon, and do all sorts of other things that none of us could ever accomplish alone.

Because the economy is an adaptive complex system, biological concepts, such as metabolism, ecology and evolution, are very useful for thinking about it. These are more than just metaphors – they contain general principles that help us understand how the economy is organized and how it works.

How concepts from biology help us understand the economy

The idea that economics could profit from concepts in biology has been around a long time.[4] In 1890, in his *Principles of Economics*, Alfred Marshall, a key founder of neoclassical economics, famously wrote, 'The Mecca of the economist lies in economic biology.'[5] Writing in the 1940s, the Austrian-American political economist Joseph Schumpeter emphasized that capitalism can be understood only as an evolutionary process of continuous innovation and creative destruction.[6] In 1982, Richard Nelson and Sidney Winter's *An Evolutionary Theory of Economic Change* made a compelling argument that the evolutionary perspective is essential for understanding how innovation drives economic growth and alters the structure of the economy over time. But Nelson and Winter were prescient when they wrote, 'We expect, however, that many of our economist colleagues will be reluctant to accept the second premise of our work – that a major reconstruction of the theoretical foundations of our discipline is a precondition for significant growth in our understanding of economic change.'[7] Ideas from biology have not, so far, led to quantitative theories making useful, falsifiable predictions. Complexity economics is starting to change this.

Adam Smith's 1776 book *The Wealth of Nations*, widely regarded as the foundation of economics, describes the economy as a complex system: 'In the lone houses and very small villages which are scattered about in so desert a country as the Highlands of Scotland,' he wrote, 'every farmer must be butcher, baker and brewer for his own family.'[8] In towns and cities, however, it was far more efficient for butchers to sell meat, brewers to make beer and bakers to bake bread than for each citizen to make all their own provisions. Trade enabled people to specialize in what each did best, and the promise of a good income motivated them not only to do things well but also to innovate. Smith understood that the economy as a whole behaves very differently from the individuals who comprise it. Stated in modern terms, he saw the economy as an emergent phenomenon.

We can restate Smith's key insight in ecological terms: The economy is organized as an ecosystem of specialists. In biology, the word *ecosystem* refers to a collection of species who interact with and affect each other. Each is a specialist, with its own unique strategy for extracting energy from the environment in order to survive and reproduce.

Species eat each other, compete with each other, cooperate with each other and collectively alter the environment. Understanding the interactions between species is essential: As mentioned earlier, if you want to understand grass, you must also think about lions. Lions protect grass: If the lions go extinct, the zebra population will rise and grass will decline. To make sense of the biosphere, it's not enough to study species in isolation; we must understand how they affect each other and alter their shared environment.

Similarly, the economy is a complex ecosystem of specialized organizations populated by workers who belong to households, whose members are consumers. In developed countries, consumption choices vary enormously – many of the products I consume are likely substantially different from those you consume. There are thousands of different kinds of firms making products, which can either be goods, like Pop Tarts or screwdrivers, or services, like legal advice and care-giving. Each firm tends to be highly specialized, because the production of most of the goods and services in the modern economy requires specific knowledge that takes a great deal of time and effort to acquire, and it is more useful and rewarding to specialize than to try doing many different things. The know-how embodied in firms and other organizations is the result of specialized occupations, each with its own unique knowledge and set of skills. Other types of organizations, such as governments and schools, play an important role in the economy. Understanding this ecosystem means thinking about the economy in terms of *networks*, which provide a universal language describing the operations of complex systems.

Economics = accounting + behavior

Networks are one of the core ideas in complex systems. For example, your extended social network describes your friends, your friends' friends and so on. You can visualize your extended social network by thinking of the people in it as *nodes*, which you can represent by dots, with *links* between friends, which you can represent by lines. If you want to be introduced to someone you don't already know, knowledge of your extended social network allows you to understand who could introduce you, or who could introduce you to someone who could introduce you. As famously shown by Stanley Milgram, we can reach almost anyone with only six degrees of separation.

Networks identify the essential building blocks of a complex system and supply a schematic view of their interactions. In a transportation network, the nodes might be cities and highways might be links. In a schematic of the financial system, the nodes might be banks and the links might be loans or investments. I say 'might' because there are usually many possible choices for nodes and links, depending on what one wants to understand. For example, the links between cities could also be based on trade, and the links between banks could be based on their common asset holdings.

The skeleton of the modern economy is a vast network of balance sheets.[9] Each balance sheet is a list of assets and liabilities, which can be physical goods and services or contracts (money, for example, is a contract). The nodes in the network correspond to organizations – households, corporations, governments or any organization that might have a balance sheet either explicitly or implicitly. Since contracts are by definition agreements between two or more parties, they link balance sheets together. My home-insurance policy is an asset on my balance sheet, but it is a liability for the company that issued it. Balance sheets are also linked by transactions, which cause goods and contracts to flow from one balance sheet to another.

The network of balance sheets underlying the modern economy is truly vast: Globally there are roughly 2 billion households and 200 million firms, as well as governments and other types of organizations that are consumers and suppliers of goods and services. Members of households are employed by firms, governments and other non-profit organizations. And households (especially in developed countries) consume thousands of different products, creating many trillions of links between households and firms; in addition, there are trillions of active contracts – so many it is difficult to count.

The global network of balance sheets is constantly in flux. Every transaction causes something to leave one balance sheet and appear on another. There is constant turnover, causing some nodes to disappear and new nodes to be added. Innovation creates new types of goods and new contracts. The network of balance sheets in the modern world is vastly more complicated and interconnected than it was 10,000 years ago.

We can think about the economy schematically as an interaction of accounting and human decision-making. Accounting is represented by the network of balance sheets, which continually changes as people make economic decisions. These decisions take many forms: What product to consume? What product to offer? Whom to hire? When to borrow? How much and from whom? All these decisions constitute economic activity. To understand the economy, we have to understand human behavior as it relates to economic decision-making, and we need to understand how this interacts with the underlying network of balance sheets.

Accounting and decision-making both pose difficult problems but in very different ways. Accounting is complicated but well-understood, whereas decision-making requires an understanding of human nature that is still incomplete.

Accounting consists of a mechanical set of rules that measure economic activity. Accounting has a central conservation law, 'equity equals assets minus liabilities', much like the conservation laws for energy, momentum and other quantities that play a central role in physics. Tracking and understanding the vast, complicated, interconnected balance sheets of the real global economy is challenging. We now have enough computer power that this is possible, but we aren't taking full advantage of this capability.

Because people don't follow simple mechanical rules, modeling human decision-making is an even bigger challenge. Understanding how people make decisions, both as individuals and in groups, is an important goal of the disciplines of psychology, sociology, anthropology and political science. As noted, the last fifty years have seen the emergence of behavioral economics, which draws on insights from other social sciences to study how we make decisions that affect balance sheets, but we still lack a comprehensive theory of economic decision-making.

The network of global balance sheets can't be studied in literal detail – it is too huge and complicated, and much of the necessary information is opaque due to confidentiality. We have to simplify the problem by dividing the network of balance sheets up into pieces and aggregating each piece. We take averages for countries or regions and study their interactions. GDP, for example, is a measure of economic activity. Macroeconomics studies the interactions of aggregate quantities, like GDP, unemployment, inflation and interest rates, within and between nations; microeconomics studies the interactions of balance sheets on a finer scale, but without trying to look at the whole economy. In Chapters 5 and 6 we will discuss how traditional economics understands the network of balance sheets from the top down, while complexity economics understands it from the bottom up.

3.

Understanding the Metabolism of Civilization

The labour process [. . .] is purposeful activity aimed at the production of use-values. It is an appropriation on what exists in nature for the requirements of man. It is the universal condition for the metabolic interaction between man and nature [. . .] common to all forms of society in which human beings live.

Karl Marx (1867)[1]

Back in the 1980s, in the early days of complex-systems science, the main focus was on the problem of self-organization.[2] How do organized configurations of matter, like life or the brain, spontaneously emerge from the disorganized background following the creation of the universe? While important and fascinating, this is a problem that will likely take centuries if not millennia to solve, and so far we have made only incremental progress. As the science of complex systems matured, it branched out to tackle easier problems with more immediate practical applications. Some of the most important early successes involved networks.[3] The pioneering sociologist Mark Granovetter introduced the concept of social networks, as discussed earlier, in which the nodes are people or firms and the links between them can take many different forms. He showed how the economic relationships between people or firms are 'embedded' in social networks, making them behave differently than they would in an abstract market setting.

Complex-systems scientists such as Mark Newman, my colleague at SFI, embraced the idea of networks and developed new mathematical methods for applications to data.[4] They showed how networks could be used to understand and improve many different real-world systems, such as how to structure connections between internet hubs to make them more reliable and less prone to sabotage, or how to allocate vaccines to ensure the maximum effect in reducing the spread of disease.

Network modeling is one of the few complex-systems ideas that has made its way into mainstream economics. Stanford economist Matthew Jackson (like me, an external faculty member of SFI) has produced a body of insightful work on the role of networks in economics, showing, for example, how social networks influence employment and inequality.[5]

Representing a system as a network begins by identifying its building blocks and their interactions: asking, 'What are the most important nodes?' and 'Are there communities? If so, what are they?' (Mathematically speaking, a community is a set of nodes that interact with each other much more strongly than they do with other nodes.) Websites provide a good example. The most important websites are not necessarily those with the most traffic or those that are linked to the largest number of other websites. Rather, they are the websites that are linked to the largest number of other *important* websites. Although this sounds circular, network theory provides a way to unravel this conundrum and identify the websites that are the most important. To see how the algorithm works, imagine a 'random walker' that starts at a random node (website) in the network. The walker then randomly jumps to one of the websites that the starting website is linked to. This process is repeated many times as the walker wanders through all the nodes. The importance of any one of these websites is proportional to the frequency with which it is visited. This algorithm, which is called PageRank, was patented by Larry Page and Sergey Brin and is used in the Google search engine.[6]

Network analysis is also useful for understanding lending between banks. In the interbank-lending network, banks are nodes and loans between banks are links. If a bank cannot pay back its lenders, this may cause the lenders to default as well, setting off a chain reaction. Suppose you're a regulator and want to keep the financial system from collapsing. Which banks are most important? If you can prop up only a few, which should you choose? The biggest banks? The banks that lend to the largest number of other banks? In fact, the answer turns out to be similar to the story for websites: The most important banks are those connected to the largest number of other important banks.[7] The same PageRank algorithm that was invented for web pages can be used to identify the banks whose collapse would cause the most harm. This makes it possible to identify the best banks to support during a financial crisis, and has been used by organizations such as the European Central Bank to identify Systemically Important Financial Institutions. That the same basic idea can be used to understand systems as different as banks and the internet (and many others) illustrates the value of complex-systems thinking. We'll say more about the important applications to the financial system in Part III.

Network theory offers a conceptual framework that helps us analyze ecologies of specialists in the economy. This approach has practical applications, like making better predictions about economic growth or understanding how quickly the job force can adapt to technological change. In this chapter, I present a static network model for production networks that was developed in mainstream economics, and discuss how my collaborators and I modified it to successfully predict economic growth. This was a key step in constructing the model that successfully predicted the economic impact of the COVID pandemic discussed in the Prologue.

The economy is like a metabolism

My introduction to networks came about because of my interest in the origin of life. My early papers on networks, in the mid 1980s, were written in collaboration with Norman Packard.[8] We developed the concept of *meta-dynamical systems* – networks that could change in response to their environment and also evolve over time. In one of the papers, we teamed up with Stuart Kauffman, a remarkably creative medical doctor and developmental biologist who joined SFI's faculty a few years later, to make a model for the prebiotic origin of life. We used networks to ask whether metabolisms might have existed even before the origin of life as we know it on the Earth now. In other words, we posited the existence of a possible simpler form of life that you might call *proto-life*, which might have been a precursor to prokaryotes, the earliest of Earth's life forms.[9]

Richard Bagley, a chemistry graduate student at UC San Diego, came to Los Alamos to work with me and wrote his thesis there, under my supervision. Rick's thesis was a model for the origin of proto-life. It built on the abstract chemical networks that Stuart, Norman and I had developed, going beyond our schematic efforts and explicitly simulating polymer nodes and the chemical reactions linking them.[10] His work showed that our hypothesis of the spontaneous emergence of a prebiotic metabolism was indeed plausible.

The simulated metabolism that Rick built is closely analogous to the way goods and services are produced in the economy. But before I can explain why this analogy holds, it's helpful to describe the way the economy produces things in network terms. Economic activity in the modern economy is highly specialized. The production network reflects the division of labor in producing goods and services. The nodes are companies that provide those goods and services, and the links are the transactions among companies or between companies and households; households can be thought of as a special kind of industry that consumes the products of the other industries and provides them with labor. Although we often take it for granted, the production network forms the backbone of the economy. It is the delocalized engine that creates the goods and services that populate and flow through the network of balance sheets.

To see how the production network functions, let's examine how

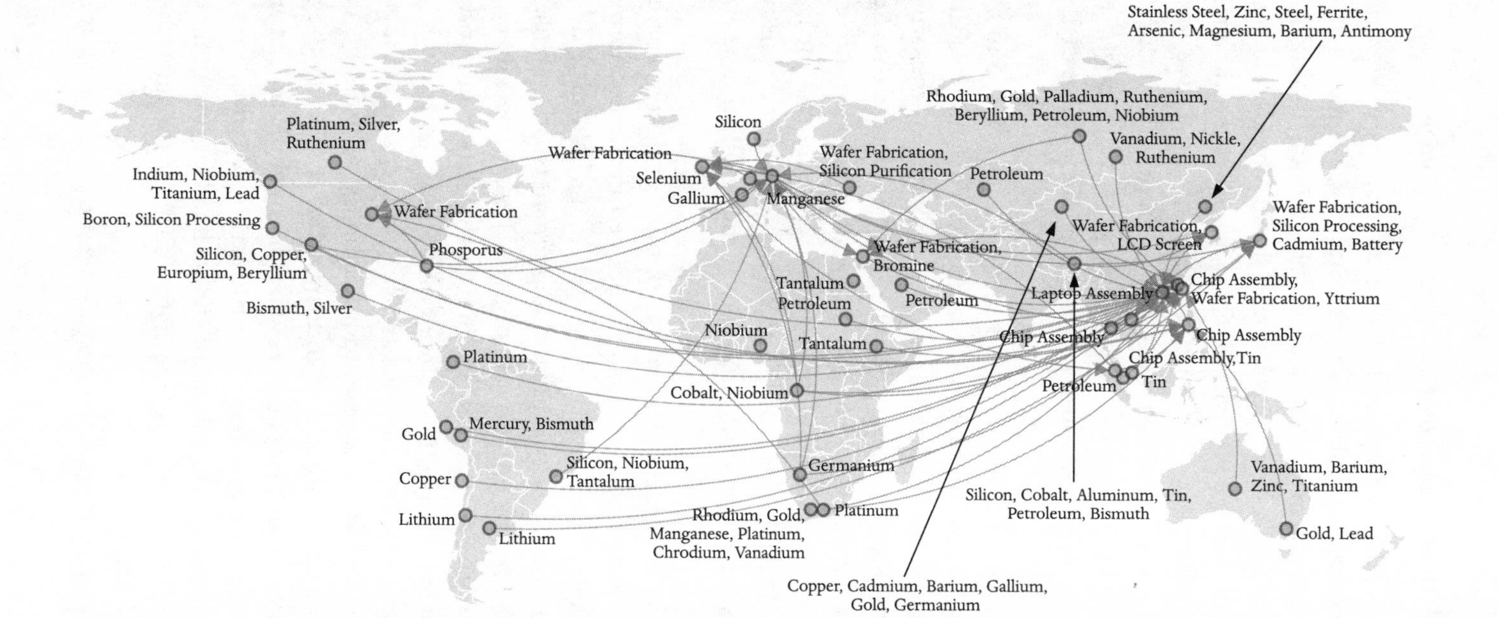

Figure 2. *Supply chain of a laptop.* The nodes correspond to products, which are shown as circles. These are connected by links that show which products are inputs to other products.

it produces a laptop. A schematic view of the supply chain of a laptop is shown in Figure 2. Many different specialized products are needed to manufacture a laptop. The nodes in the network are these products and the links connect each product to its constituents.

To produce a laptop, raw materials are extracted from every continent except Antarctica, transported to other locations and processed to make composite materials. These in turn are used to make the individual parts that are then combined and assembled elsewhere to create the laptop. The process involves many firms in different industries. Each industry makes its product and sells it to another industry further downstream. The final product, the laptop, is both a consumer good and an intermediate item used to make other products: Engineers use laptops to design many things, including new laptops. This is one of many loops in the production network. Though we often refer to pieces of the production network as *supply chains*, this is a bad metaphor: The production network is full of branches and is more like a tangled web than a chain.

The physical inputs for laptop production are only part of the story. A laptop is designed by engineers, who work for companies that get started by raising capital, which comes from investors, including bankers and other financiers. Once it's made, a laptop must be transported, distributed and marketed. The engineers, fabrication facilities and computer stores are all housed in buildings, so we must include the construction industry among the inputs. The buildings sit on land, so we must include real-estate companies. Manufacturing requires energy, so we have to include that, too, not to mention the janitors, cooks in the company cafeterias, accountants and lawyers, the occasional plumber to unclog a toilet – the list is long. Every industry is connected to many other industries, and the connections loop back on themselves: Accountants, real-estate agents and engineers all use laptops to do their jobs.

The making of a laptop illustrates how the economy operates like a big metabolism. The production network has a structure similar to the proto-life model described earlier. Their definitions are the same except for a change in noun phrases:

> A *metabolism* is a network of *chemical reactions* that transforms a *food set* of *chemical inputs* into other *chemicals* that provide *energy and raw materials for living cells*.
>
> A *production network* is a network of *firms* that transforms *natural resources and labor* into *goods and services for consumers*.

The economy coordinates the labor of billions of people, so that each network component is available when and where it is needed.

There is an essential difference between production networks and biological metabolisms: In biology, metabolisms are localized within each organism and ecologies are built from the interactions between organisms. But the economy is structured as if it were one big superorganism, with a single metabolism. The metabolism of this superorganism operates through the actions of an ecology of specialists, each of whom the metabolism supports. This remarkable circular process thrives on the specialization of its goods and services and the specialization of the labor required to produce them.

An early pioneer of complexity economics

By 1925, post-revolutionary Russia was becoming increasingly dangerous for Wassily Wassilyevich Leontief. A man from a wealthy family, he'd already been arrested several times by the Cheka secret-police because of his campaigning in support of academic autonomy and freedom of speech. With the net closing ever tighter, he feigned an untreatable terminal illness that led the authorities to take pity on him and allow him to spend his final months in the West. Having escaped the Soviet Union, he eventually made his way to Harvard where he carried out groundbreaking work on production networks.

Leontif took seriously Adam Smith's observation that specialization is an essential feature of an economy and made a mathematical model for how individual industries in specialized roles interact to create economic output.[11] He was one of the first economists to embrace the digital age: In 1949 he used an early computer, the Harvard Mark II, to solve a system of linear input-output equations that divided the economy into 500 industries. This was quite a feat at the time. (Since then, the field has in a sense gone backwards – typical models now only use between 40 and 100 industries.) His systems of equations were effectively network models of the economy. Though no one used the term while he was alive, and he was part of the mainstream, I consider Leontief a pioneer of complexity economics.

The nodes of Leontief's production network are industries, comprising different firms that make a family of products (like automobiles) that can be lumped together and treated as if they were a single product. The products can be physical goods (like screws) or services (like pizza delivery). Each industry buys products from other industries, converts them to make a new product, and sells them, in turn, to households or other industries. The links between the nodes are the inputs required to produce each product. Money flows in the opposite direction to products: Physical goods flow from suppliers to customers; customers pay their suppliers.[12]

Leontief's model assumes that each product is produced via a 'recipe' (called a *production function* in economics) that specifies how much of each input is needed to make the output. This can be measured in monetary terms as the value of the products that companies buy and sell. Leontief's original model makes some strong assumptions, such as setting markups (the difference between revenues and costs) to zero and assuming that the economy exists in a steady-state equilibrium. Neither of these assumptions is true, but they provide a crude approximation that makes it possible to solve the resulting equations for the prices of all the products. These equations closely resemble those that Rick and I used (and that biochemists use) to understand metabolisms.

Given his status as a defector, it is ironic that the Soviets were perhaps the biggest beneficiaries of Leontief's equations, which they used to make the pieces of a planned economy fit together. Leontief's equations solve the chicken-and-egg problem caused by the many positive-feedback loops in the production network. For example, the Soviets desperately wanted to increase steel production. To do this, they had to build up the industries that provide inputs to the steel industry, such as iron, coal and railways. But building railways requires steel, as do the machines for extracting iron and coal. Thus, to increase steel production, you need to increase steel production. Leontief's methods helped solve this conundrum and allowed the Soviet planners to invest in each industry so as to increase production as quickly as possible.[13]

In economics, models of the production network with multiple industrial sectors are usually called *input-output models*. The mainstream interest in input-output models has ebbed and flowed over the last seventy years. For two decades following Leontief's pioneering work, there was a major effort to predict the behavior of the economy using input-output models with hundreds of industrial sectors. This was not very successful, and, as we will discuss in Chapter 5, in the 1970s macroeconomics took off in a very different direction. Detailed models of the production network were abandoned, and by the turn of the millennium they had become a footnote in macroeconomics textbooks.

Meanwhile, in the 1990s, my colleague H. Eugene Stanley from Boston University and some other physicists started a new movement they called *econophysics*, promoting a physics-style approach to understanding the financial system and the economy. One of the early econophysics papers from Gene's group pointed out that the year-to-year fluctuations in the growth of large companies behave in a very peculiar manner.[14] The 'law of large numbers' suggests that fluctuations in output should average out when we consider the economy as a whole, but this is not what happens – the fluctuations are much larger than one would normally expect. The French-American economist Xavier Gabaix postulated that this is due to the extreme variation in the size of companies, and he used input-output models to develop his theory.[15] Gabaix's work led to a rekindling of interest in input-output models, which have now branched out in other directions.

Static models of production networks are one of the branches of economics where mainstream economics and complexity economics overlap. The model I am about to describe has much in common with models that are currently popular in the mainstream. Nonetheless, because my colleagues and I started in a different place, taking our inspiration from ecology, we came at the problem in a different way and found ways to apply Leontief's model that mainstream economists missed.

The ecology of production networks

The second graduate student who played an important role in my work on production networks is James McNerney. James, a physics student, wanted to work on something relating to economics and was particularly interested in the economics of climate change. He came from Boston University to the Santa Fe Institute in 2008 and spent several years there working with me. Ecology was always well represented at SFI, and this led James and me to think about whether we could apply concepts from ecology to economic production networks.

We were particularly interested in how technological improvement in one product might cascade through the production network and result in technological improvements in other products. James extended Leontief's equations by adding a simple model for technological progress, assuming that the net result of progress is that the same output is made with less of each input. The links in the production network remain the same, but their weight decreases in response to each industry's progress, and as inputs go down, so do product prices.[16] (In reality technological progress has many other effects, but it is often useful to start simple.)

Our idea was based on the ecological concept of a food chain. Imagine a simplified ecology consisting of lions, zebras and grass, and assign each species a trophic level – a numerical rating based on their place in the food chain. Since grass gets its energy directly from the Sun and the atmosphere, it is usually assigned a trophic level of 1. Zebras eat grass, so they get a trophic level of 2. Lions eat zebras, so they get a trophic level of 3.

The real world is more complicated. Organisms like humans and pigs are omnivorous, eating at multiple levels in the food chain, and decomposers (like worms eating a lion's corpse) create loops. In reality, the food chain is actually a food network. Nonetheless, it's possible to assign trophic levels, even in such messy situations, through a set of equations stating that the trophic level of each species in an ecology is one more than the weighted average of the trophic levels of the things it eats. The resulting trophic levels are no longer whole numbers, but they're still helpful in understanding how a disturbance – for example, the removal of a top predator by overfishing – affects other species.

We saw that we could compute trophic levels for industries in the same way. The production network is analogous to a food web. In a food web, the nodes are species and the links correspond to what they eat; in the production network, the nodes are industries and the links correspond to their inputs. While households can be viewed as an 'industry', they play a special role – they are the end-users of the economy.[17] James and I made households the reference point (like the Sun in a food web) and assigned them a trophic level of zero.

In economics, the trophic level of an industry can be thought of as the depth of its supply chain. It measures the number of steps required to make an industry's product.[18] An industry whose only input is labor has a trophic level of 1, because its only input comes from households, which by definition have a trophic level of zero. An industry with inputs from industries other than labor automatically has a trophic level greater than 1. An industry has a high trophic level if the fraction of its inputs from labor are small and if its other inputs have high trophic levels. Service industries tend to have low trophic levels and manufacturing industries tend to have high trophic levels. (Note that the trophic level is essentially the same as what is called an *output multiplier* in economics, though we used the concept here differently than in earlier work.[19])

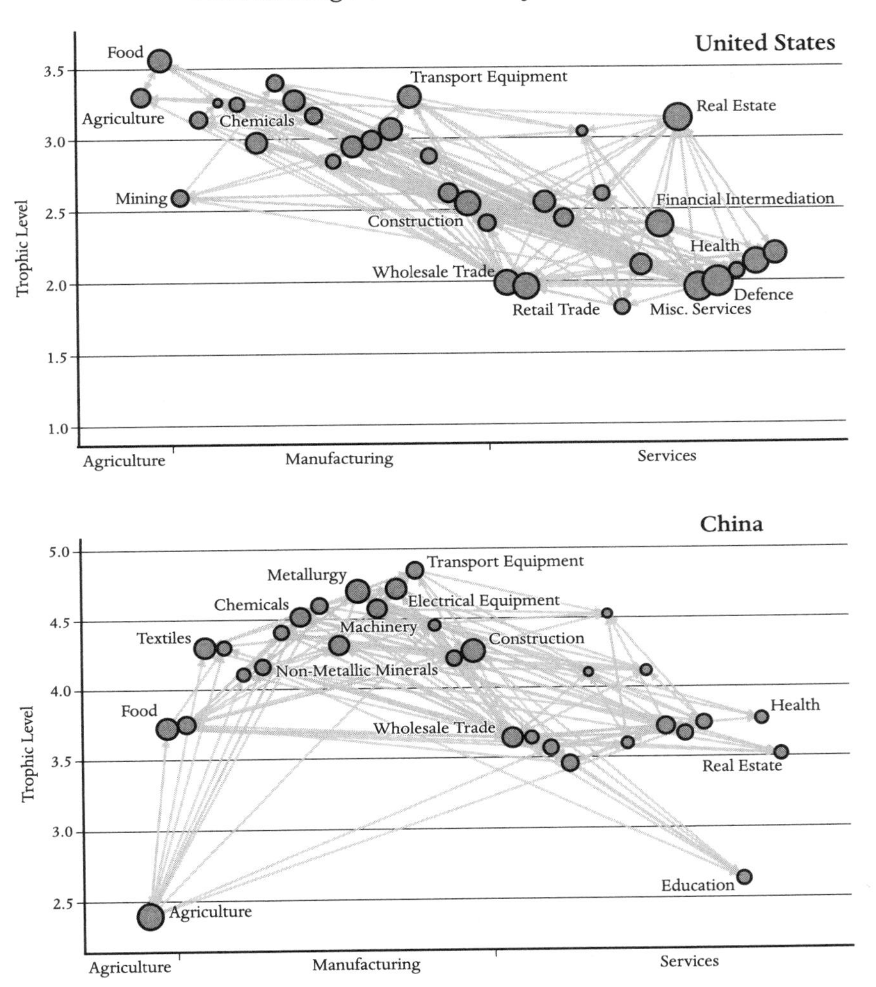

Figure 3. *Comparison of the trophic structure of the US and Chinese economies.* The industries are arranged horizontally based on industry type. The vertical axis shows the trophic level for each industry. The size of the circles (nodes) corresponds to the revenues of each industry. The arrows (links) correspond to the largest inputs for each industry.

Figure 3 compares the production networks for the United States and Chinese economies. The trophic level of each industry is plotted along the vertical axis, while the horizontal axis places industries according to a standard industrial classification scheme. Plotting the data this way makes some key differences between the two economies obvious at a glance.

One of the most striking differences evident in Figure 3 is that on average the trophic levels of Chinese industries are higher than those of US industries. There are two reasons for this. One is that the Chinese economy is much more heavily concentrated in manufacturing industries. In China, manufacturing industries such as Electrical Equipment or Metallurgy play a big role in the economy; they tend to have high trophic levels because there are many steps in their production. In contrast, the United States concentrates heavily on industries with shallow supply chains such as Miscellaneous Services or Health.

The second reason trophic levels of Chinese industries are higher is that they pay their workers lower wages. Since labor is a smaller fraction of the cost of the inputs, the contribution of the household sector is lower. This means that the relative amount paid for manufactured inputs is higher, which raises the trophic level of industries in China.

It is also interesting to compare specific industries: American agriculture, for example, has a high trophic level, reflecting a high degree of mechanization. The US industry with the highest trophic level is Food, because Agriculture, by far its biggest input, already has a high trophic level. In China, Agriculture is highly labor intensive, giving it a lower trophic level, like that of service industries.

But what does trophic level have to do with technological innovation and economic growth? Economic growth depends on technological innovation. What James realized is that, since innovations are passed along the supply chain, products made by industries with high trophic levels will tend to improve faster than products made by industries with low trophic levels. Think about laptops: The cost of making a laptop depends on the whole industrial ecosystem that supports it from below. If the tantalum-mining industry becomes more efficient, this will drive the cost of tantalum down, so memory-chip manufacturers pay less for tantalum, which lowers the cost of memory chips. If the manufacturers improve the chip-manufacturing process, this lowers costs even more, and so on, all the way up the supply chain. We showed that this causes an amplification effect: Improvements combine multiplicatively as one moves up the supply chain. In contrast, for a service-intensive industry whose only input is labor, there are no improvements passed on from the rest of the economy. Any improvements must come from that industry itself – the only way to innovate is to use labor more efficiently.

The typical industry improves more due to the effort of its suppliers than it does due to its own effort. James showed that 65 per cent of improvements in a given industry come from other industries – or, alternatively, only 35 per cent of the improvements in a typical industry come from within the industry itself. Thus, if all industries innovate at the same rate, and if all else is equal, industries with a higher trophic level will improve faster than industries with a lower trophic level. This is because an industry with a deeper supply chain has more industries below it to provide it with innovations.

We tested our hypothesis on data for thirty-five industries in forty countries from 1995 to 2009. This database computed a price index for the products made by each industry in a given country, which gave us an overall measure of improvement based on lower prices to buy the same thing, adjusted for changes in quality. For example, if laptops get faster and have more memory but sell for the same price, the price index will drop, to account for the improved quality. When the price index drops, this means that the product has either become cheaper, or better, or some combination of the two, but in any case it signifies the overall level of improvement.

We tested our hypothesis by forecasting changes in the price index, taking advantage of the fact that trophic levels change very slowly – so slowly that they can be treated as constant over the fourteen years between 1995 and 2009. We showed that the trophic level of a given industry in a given country in 1995 predicts how much its products will have improved fourteen years later. Since there are other factors influencing improvement, the individual predictions are by no means perfect, but they worked extremely well: The average improvement for a given industry in a given country increases proportional to trophic level, as we predicted, and our hypothesis is strongly confirmed with extraordinarily high levels of statistical significance. This helps explain some well-established economic facts, such as why manufacturing industries tend to improve faster than service industries, as originally noted by NYU economist William Baumol.[20]

Trophic levels also predict the growth rates of countries. Let's call the average trophic level of the industries of a given country the trophic level of that country. According to our model, the growth in GDP of a country should be proportional to its trophic level. (As before, it is necessary to add 'if all else is equal', because there are many other factors that influence GDP growth.) When we tested our hypothesis at the country level on the same data set, countries with higher trophic levels in 1995 tended to have higher GDP growth over the next fourteen years, just as predicted. From 1995 to 2009, US GDP grew by about 1.5 per cent annually, while Chinese GDP grew by 8.7 per cent annually. How much of this difference can be explained by trophic level? According to our model, about 30 per cent of China's growth advantage over the US between 1995 and 2009 is due to its higher trophic level in the global economy, and the rest is explained by other factors.[21]

Thinking in terms of trophic levels leads to an interesting hypothesis about development paths. Although trophic levels change slowly, over many decades, they change in economically important ways. In his PhD thesis, one of my other students, Charles Savoie, estimated how the trophic level of the US economy had changed over time from 1790 to the present.[22] Bearing in mind that data from earlier years is limited and unreliable, the story goes like this: In 1790, the trophic level of the US economy was about 3.2, with trophic level and per-capita GDP roughly equivalent to that of present-day Bulgaria. By 1900, the trophic level had increased to about 4 as the manufacturing sector grew. The trophic level then remained more or less constant until roughly 1950, while per-capita GDP grew, and then began to steadily decline to its current value of 2.5, as manufacturing was increasingly replaced by services.

Will modern-day developing countries on the industrial growth path evolve in the same way? China's current per-capita GDP is a bit lower and its trophic level a bit higher than those of the United States in 1900. At present, China's high trophic level helps the Chinese economy grow rapidly, but if it follows in our footsteps, GDP will eventually grow more slowly as the country's trophic level drops due to higher wages and the replacement of manufacturing by services.

One of the most striking aspects of our predictions is that they don't depend on human behavior, at least directly.[23] The trophic level is a *structural property of the economy*. Our theory assumes the structure of the network is given. Because our predictions are independent of any assumptions about decision-making, they are robust.[24] In contrast, standard models require an assumption that firms maximize their utility, which must take a particular mathematical form, and assume a particular mathematical formula for the production function.[25]

Adam Smith argued that specialization makes the economy work better. Our work here shows that *specialization also boosts the rate of innovation in the economy*. This happens automatically: Each industry in the production network can count on the rest of the economy to improve its inputs and provide it with new technologies that will allow improvements in its own processes. Improvements in an industry with a higher trophic level automatically happen faster, even if innovations in that industry happen at the same rate as in other industries.

Our extension of Leontief's model assumes that the rate of improvement for each technology is constant in time, so the result is, like his, a static model. In the next chapter, we'll discuss the development of a dynamic, agent-based model, using principles very different from those that govern equilibrium-based models. (An example was already given in the Prologue.)

The evolution of the ecology of work

In 2012, I was invited by Eric Beinhocker to come to the University of Oxford. Eric had been a regular visitor at the Santa Fe Institute, and his 2007 book, *The Origin of Wealth,* described the problems of mainstream economics and included a discussion of work by me and others in complexity economics.[26] He had received a grant from George Soros to start a research group that would take a new approach to economics, and he wanted me to join him and lead a program on complexity economics. My original appointment was as Director of Complexity Economics at the Institute for New Economic Thinking at the Oxford Martin School, which is a cross-disciplinary organization devoted to funding science to help solve the big problems of society. But it's not an academic department, and before I could be appointed I had to become attached to a university department. Somehow, I ended up in Oxford's mathematics department, even though I've never published anything in the field of mathematics, but I have since moved to the Smith School for Enterprise and the Environment, where my title is the Baillie Gifford Professor of Complex Systems Science. I eventually got an offer to be a Senior Research Fellow in Christ Church College, where one of the fun perks is wearing a black gown and sitting at 'high table' in the dining hall that the Harry Potter movies used as a template. Oxford is a long way from Santa Fe, in many respects.

Because my position at the Oxford Martin School is supported with outside funding, I don't have teaching responsibilities, which gives me more time for research. However, I do advise graduate students, which suits me perfectly. Oxford attracts excellent students from all over the world; I enjoy collaborating with them, and my productivity has soared as a result. Since my arrival, I have been steadily building up my group, and I now have about fifteen graduate students from a variety of disciplines. Most of them approached me because they wanted to do economics but weren't happy with the approach of traditional economics departments.

The work that two of my best students, Penny Mealy and Maria del Rio-Chanona, have done on understanding labor transitions provides a good example of complexity economics. Penny, who now works at the World Bank, was my graduate student and post-doc at Oxford, where she was based in the geography department.

Maria originally studied physics at Universidad Nacional Autónoma de Mexico, but moved to mathematics when she came to Oxford to do her graduate work with me.

In the coming decades, we will experience two major economic transitions – automation (including artificial intelligence) and the green-energy transition – which are unfolding in parallel. In addition to changing the way we produce things, they are also causing seismic shifts in the workplace that will dramatically alter the occupational landscape and determine the future of work for the next generation. We have developed economic models to help navigate these transitions.

Consider the dilemma faced by coal miners. As coal mines close, miners need to find new jobs. You might think the simple solution would be to train them to operate solar farms, but the best sites for solar farms are usually far from coal mines, and the required job skills are different, so this seemingly easy solution doesn't make sense. Instead, data on labor transitions tells us that coal miners are more likely to become construction-equipment operators or truck drivers, and workers for solar farms are more likely to come from power plants or structural iron and steel mills. As these transitions happen, they alter the supply and demand for labor in each profession, causing changes that propagate through the workforce. Because everything depends on everything else, nothing can be understood in isolation. To understand how the occupational-labor network will evolve with automation and the green-energy transition, we need a more holistic view.

We have built an agent-based model that tracks the supply and demand for labor within individual occupations, showing how the labor ecosystem adapts to changes in the economy. The model helps answer questions such as, 'Which occupations will be most affected?' or 'How much unemployment will this cause?' or 'What kind of workforce training would be most effective?' It also suggests how the lack of skilled labor for particular industries might affect production. For example, to minimize climate change, we need to make the green-energy transition as fast as possible. Can the workforce adapt quickly enough without slowing down the green-energy transition? Penny managed an initial effort to get the lay of the land, and Maria headed the model's development.

We began by taking snapshots in time, constructing a static network in which the nodes were occupations and the links were relationships between them. The links were based on the similarity of work activities – designing, say, or recruiting and hiring. This network, which we call the *job space*, is shown in Figure 4.[27] The nodes

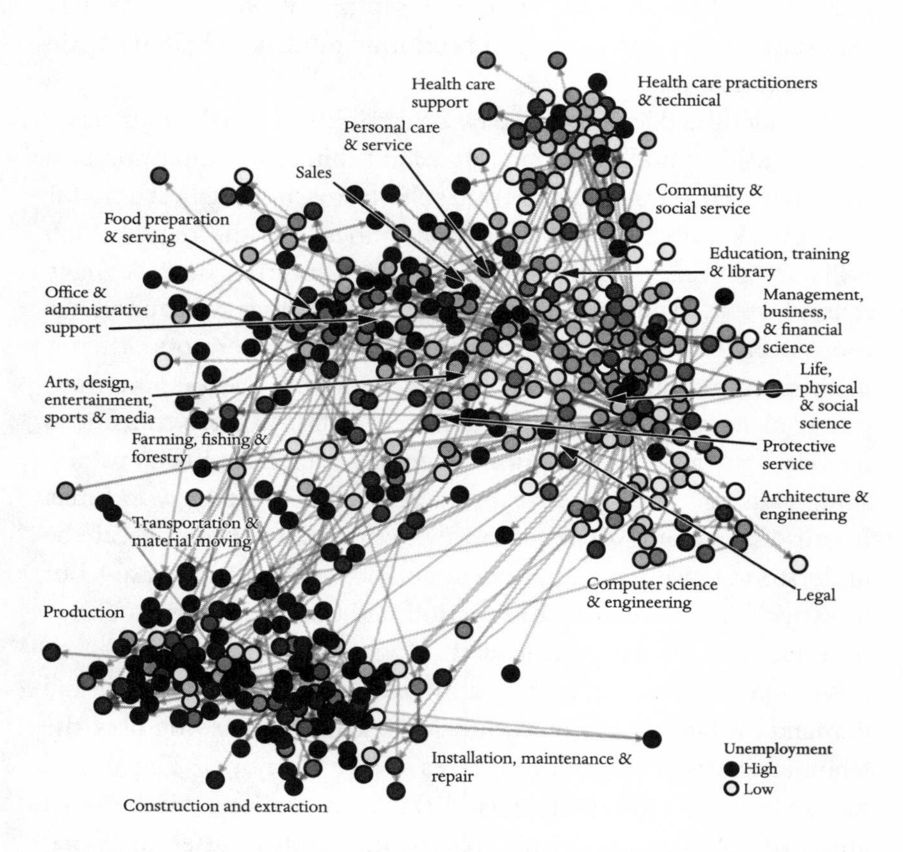

Figure 4. *A network view of the job space for the United States in 2016.* Nodes (circles) are occupations, and links are for changes of occupations that occur frequently. Black indicates an occupation with high unemployment, white indicates low unemployment, with shades of grey in between. The labels indicate broad occupational categories.

are arranged so that occupations that share many work activities are close together. The similarity of work activities predicts the likelihood of job transitions between occupations better than methods previously proposed.

In Figure 4, the nodes in the job space are shaded based on the average unemployment associated with each occupation. This makes some patterns self-evident: For example, occupations in the technology or health care parts of the space have low unemployment rates, while jobs in the construction and mining part of the space have high unemployment rates. The diversity is striking: The unemployment rate for nurses is 0.1 per cent; for boilermakers it's 25 per cent. The clustering of unemployment in the job space occurs because occupations that are close to each other share work activities, which means the supply and demand for their labor tend to be similar.

Static network plots provide a useful snapshot at any given point in time, but we wanted to be able to predict how the landscape would change. To do this, Maria built an agent-based model for the labor market.[28] It simulates all the steps that workers take when they look for jobs. We start with a static network, but the model is dynamic: As before, each node represents an occupation, but the model also includes information attached to each node that describes the state of each occupation – number of workers, job vacancies – and calculates what happens month by month as workers are fired and new job vacancies are created based on that occupation's current labor demand. Links are based on the frequency with which workers have changed occupations in the past.[29] Workers apply for new jobs both in their current occupation and in other occupations. The relative rate that they apply for jobs in new occupations is based on data from the US census. To keep things simple, we assume that all workers applying for a position are equally likely to be hired. Workers in occupations with little competition find new employment quickly, but, if competition is stronger, finding a new job may take many months or even years. Whenever there are changes in labor demand in our model, the effects ripple through the job space, as each occupation affects its neighbors. The beauty of the dynamic model is that it allows us to keep track of interactions among occupations across the entire job space and understand how everything affects everything else.

To study how automation will affect occupations in the job space, we used predictions by our Oxford colleagues Carl Frey and Mike Osborne on the types of jobs that computerization will eliminate.[30] It's important to keep in mind that, historically, automation has not resulted in higher unemployment; instead, it has shifted employment from one occupation to another. For simplicity, we assumed that the Frey and Osborne predictions changed the relative demand for occupational labor without affecting the total demand for labor. We took their predictions as given and then studied how workers migrate across the job space as old jobs are eliminated and new jobs are created.

The simulations of the occupational-mobility model indicate that the transition will be far from uniform: There are regions of the network where workers easily find new jobs and others where jobless workers get trapped because there are no suitable alternatives, causing a substantial and persistent boost to average unemployment. In the model, average unemployment rises from about 5.3 per cent before the automation shock to 6.7 per cent at the peak of the transition. We compared our results to those based on a hypothetical 'complete network', which assumes workers can freely transition from any occupation to any other. In the hypothetical complete network, the unemployment bump is much smaller than our more realistic model predicts, starting at 4.1 per cent before the shock and peaking at only 4.7 per cent. This difference in predicted unemployment rates makes it clear that limits on occupational mobility in the real world amplify overall unemployment and help determine how long shifts in unemployment will last during a transition.

Our occupational-transition model predicts some counterintuitive effects. Dispatchers (like people who coordinate taxis[31]) and pharmacy aides both face the same high (72 per cent) probability of automation in their jobs, but the shifts in their long-term

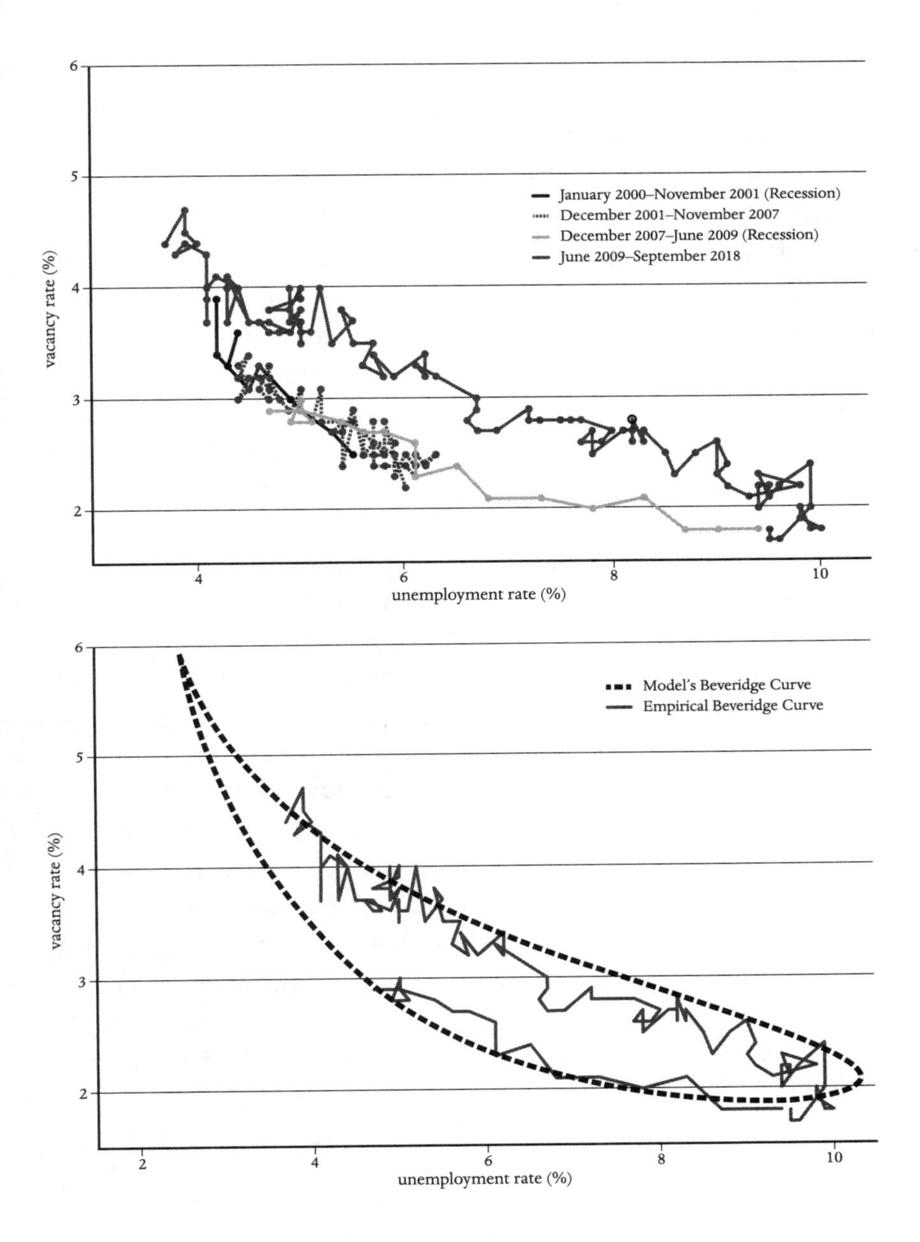

Figure 5. *The Beveridge curve* shows the movements of the job-vacancy rate against the unemployment rate at different times. The top panel shows the Beveridge curve for the US from January 2000 to September 2018. The bottom panel superimposes the observed Beveridge curve with that predicted by our model.

unemployment rates are quite different. The automation shock causes a 19 per cent increase in the long-term unemployment for dispatchers, while the pharmacy aides' long-term unemployment *decreases* by roughly the same ratio. This is because, after the transition, there is more competition with workers from other occupations seeking jobs as dispatchers than for pharmacy aides (presumably because dispatchers require less training). The influx of new workers from other professions seeking jobs makes the job market for dispatchers more competitive. Similarly, even though child-care workers are directly unaffected by automation, their long-term unemployment *increases* by 6 per cent, because workers from many occupations undergoing automation can easily become child-care workers. These effects make sense once you think about them, but to get their magnitudes right requires a model like ours that takes all occupations into account at once.

We also got an unexpected bonus: Even though we didn't build it to do this, our model provides a quantitative explanation for an important economic phenomenon called the Beveridge curve, which describes one of the ways in which it's easier to fall into a recession than to recover from it. The Beveridge curve plots the job-vacancy rate against the unemployment rate. As we enter a recession, job vacancies go down and unemployment goes up, and when we recover from a recession, the opposite happens. However, there's a difference between the trip down and the trip up: At the same level of vacancies, the unemployment rate as we enter a recession is lower than it is as we recover from the recession. For example, as the US economy collapsed in December 2007, the vacancy rate was about 3 per cent and the unemployment rate was about 5 per cent. The vacancy rate kept falling, dropping below 2 per cent. When the vacancy rate began to rise again in the recovery phase, increasing to the 3 per cent level by March 2014, the unemployment rate remained at 6.7 per cent. Because of this difference, the path when we fall into a recession is different than the recovery path, and a plot of vacancies vs unemployment for the 2008 recession looks a bit like a teardrop, as shown in Figure 5.

The Beveridge curve is a good example of an emergent phenomenon: We didn't build the model to understand it, so we were pleasantly surprised when we discovered that our model could explain it. In fact, Figure 5 shows that the Beveridge curve that our model makes when we simulate the business cycle for the 2008 financial crisis has a similar size and shape, and the same direction of movement, as the real thing. This is encouraging because a good model should explain as many different phenomena as possible. In sharp contrast to the mainstream work on this subject, which assumes the Beveridge curve is an equilibrium phenomenon, our model predicts that it is a *disequilibrium* phenomenon – that is, a transient response to changing conditions in the economy. According to our model, more sudden recessions should cause 'fatter' Beveridge curves or, in other words, a bigger gap between the path up and the path down. We are in the process of testing our predictions empirically by looking at past recessions. In contrast to standard theories, our model predicts that the area inside the Beveridge curve is proportional to the suddenness of the recession.

There are also practical policy applications. To recover from a recession quickly, one would like to make the Beveridge curve as 'thin' as possible: For a given level of job vacancies, one would like to make unemployment on the trip up (the recovery) as close as possible to where it was on the trip down. Our model provides insight into how to accomplish this. For example, if we artificially make it easier to move between occupations, the Beveridge curve gets thinner.

Our occupational-transition model is a work in progress, and we are just beginning to explore all the things it can do, but we believe it will ultimately allow a more quantitative understanding of the dynamics of occupational unemployment. At a policy level, it could increase the effectiveness of job-retraining efforts by targeting the right occupations and reducing the problems created by shifting demand for labor across different occupations in response to factors like automation. We are designing a new version of this model using detailed microdata from Brazil to calibrate it more precisely. To study the green-energy transition, which is happening in parallel with automation, we are also developing a model for how the production network will shift as we move from fossil fuels to renewables. We are then using the production-network model discussed earlier in tandem with the occupational-transition model to predict how the job landscape will be altered by the transition and whether bottlenecks are likely to develop that might slow the green-energy transition down. We are also incorporating geographical, as well as occupational, mobility. This will allow us to anticipate such bottlenecks, so that policymakers can plan ahead and focus retraining efforts on the occupations that will be most needed to keep the green-energy transition on track.

4.
How Simulations Help us Understand the Economy

For most problems found in mathematics textbooks, mathematical reasoning is quite useful. But how often do people find textbook problems in real life?

Lynn Steen[1]

What is a simulation?

Computers allow us to mimic just about anything. Boeing creates a full-blown virtual-reality simulacrum of every plane it designs. Pilots fly simulated planes whose controls and responses look and feel much like those of a real plane, and virtual flight attendants walk down the aisles and serve drinks to simulated passengers to make sure the seats and trays are well placed. Although there are still limits to their capabilities, computer simulations now enable us to create digital twins of countless real-world systems, and they play an important role in almost every branch of science, including the social sciences – except for mainstream economics!

A simulation is often preceded by building a network model. You can think of the network as the skeleton, and the simulation as the living tissue filling out the skeleton and enabling the body to function. A simulation goes beyond a network model by capturing the dynamics: How do the nodes and links, and their properties, change over time? Simulation is a bottom-up approach, acting at the level of the individual building blocks, letting the collective behavior of the system emerge. By definition, emergent behavior is bottom-up, and simulation is often the only method for understanding it.

Simulation in the social sciences is more challenging than it is in the physical sciences. Atoms don't think or make choices, while stars obey the laws of gravity, which are precise, inviolable and well understood. But there are no laws that can reliably tell us how people will behave in any possible situation. People have agency; they make decisions, which are much more difficult to predict than physical processes. There is now a huge body of work in psychology and sociology that helps us understand how people behave; nonetheless, our models for human behavior are still imprecise and controversial, and, most important, they don't cover all aspects of economic decision-making.

Agent-based models have been very successful for problems where decision-making is reasonably simple. A good example is traffic modeling, in which drivers are the agents.[2] In a traffic simulation, each simulated driver interacts with simulated drivers in nearby cars, speeding up and slowing down as needed. Traffic jams are a simple example of an emergent phenomenon: As soon as one driver slows down on a crowded road, those behind must slow down too. Cars begin to pile up, which causes even more slowing down, until the traffic comes to a near-halt. Agent-based models do a good job of predicting the conditions that can lead to traffic jams, and they help city planners design transportation systems. Some cities, such as Portland, Oregon, have built hyper-realistic simulations, with a one-to-one representation of every road, building and almost every car, and there are now several competing companies that provide traffic simulation for hire.[3] These simulations accurately predict travel times under different conditions, allowing planners to ask 'what-if' questions and experiment with adding or closing lanes or building bridges to see how such changes would affect traffic. One of the reasons traffic modeling works well is that the decision-making isn't very complicated and is easily captured by simple rules of thumb, involving factors like the typical minimum spacing between cars and the typical reaction time for braking.

Agent-based models have also become essential tools in epidemiology. Epidemic simulations track the movements and interactions of individual people, who can be healthy, infected or immune. Such models can incorporate transportation routes and flight schedules connecting cities and the locations of hospitals and schools. Simulations can investigate 'what-if' scenarios and, in some cases, they have accurately forecast the likely effect of vaccination protocols or other public-health measures, helping to achieve the best possible outcomes with the least effort.

The COVID pandemic presented a challenge to epidemic simulations for two reasons. At the onset, there was a great deal of uncertainty about how the virus was transmitted and whether or not it could be transmitted by people who were asymptomatic. Disease propagation depends crucially on these parameters, which threw off the predictions of the models in the early stages of the pandemic. As data poured in, these parameters became more certain, but another problem appeared: People began to change their behavior. In some places, people were careful about wearing masks and following social-distancing protocols, and in other places they weren't. As time wore on, in some places there was a backlash against masks and vaccinations, sometimes fueled by politics – circumstances difficult to predict in advance. Despite these problems, the best models provided useful predictions that helped policymakers make better decisions.[4]

Simulation models are never perfect, but they can be good enough to be useful.[5] We'll discuss the controversies about models based on human behavior in Chapters 5 and 6. But, before we delve into that, we need to understand an essential advantage of agent-based models, which is the graceful and effortless way they incorporate nonlinearity.

Nonlinearity, emergence and the need for simulation

Nonlinearity is a necessary condition for emergence; without nonlinearity, there would be no emergence and thus no such thing as a complex system. Nonlinearity underpins almost all the big, interesting open questions – such as the origins of intelligence, consciousness and life – as well as the recent huge advances in machine learning. It also plays important roles in the economy – roles that have yet to be properly explored and can be fully understood only through simulation.

By definition, a system is nonlinear if the whole is *different* from the sum of its parts. In contrast, a system is linear if the whole is *equal* to the sum of its parts. Nonlinearity is the norm in the real world and economics is no exception. For example, as already mentioned, production functions describe how much a firm or industry can produce with a given set of inputs. A production function is like a basic recipe with a list of possible variations. In some cases a single input may be absolutely essential, or in other cases substitutions may be allowed. The amount that is produced is not simply the sum of the inputs, but rather depends on the inputs in a more complicated way, which means that the formula relating the inputs to the outputs is a nonlinear function.[6] In Chapter 7 we'll see how the nonlinearities of the production function can give rise to oscillations in the economy that look like business cycles, and in Chapter 11 we'll see how the nonlinear behavior of loan repayment can cause the financial system to crash.

It is often tempting to avoid nonlinearities because they complicate mathematical analysis. Economists greatly prefer equations with what are called *closed-form solutions* – that is, formulas that can be used to simply plug in numbers and get the answer. All scientists like closed-form solutions, because they're convenient and easy to understand. But the quest for closed-form solutions often distracts economists from solving the most important problems. Economists love mathematics and eschew simulations. Our understanding of economics has suffered as a result.

Before the advent of the computer, the mathematical arsenal for scientists was limited to what they could solve with pencil and paper. This put an enormous premium on finding closed-form solutions. Because linear equations always have closed-form solutions and nonlinear equations rarely do, expedience dictated that almost all models in science were based on linear mathematics. The world of nonlinear mathematics remained largely unexplored.

When I was a graduate student at UC Santa Cruz, in 1975, I was inspired by a lecture by Stan Ulam, one of the greatest twentieth-century mathematicians.[7] He began by apologizing for the title of his lecture, which was 'Nonlinear Mathematics'. Since almost everything is nonlinear, he said, calling mathematics *nonlinear* was like referring to 'nonelephant' animals. To illustrate his point, Ulam told a standard joke about a mathematician who is found searching for something under a lamppost. A passerby asks, 'What did you lose?' and he replies, 'My keys.' Trying to be helpful, the passerby responds, 'Can I help you look for them?' The embarrassed mathematician responds, 'Well, actually, I didn't drop them here, I dropped them over there in the dark somewhere. But at least I can see here.'

Until digital computers arrived, most of science suffered because we couldn't solve most of the equations we needed to solve to understand the world. All we could do was look under lampposts. The linear ground under the lampposts got thoroughly explored, while the nonlinear areas outside the circle of available light were neglected. Ulam was one of the earliest mathematicians to use computers to explore nonlinear mathematics, and one of the first to demonstrate emergent behavior in complex systems.[8]

The fact that almost all equations are nonlinear doesn't mean that almost all equations lead to emergent phenomena. The presence of nonlinearity *per se* doesn't imply emergent behavior. In fact, usually it doesn't. Complex systems are special. We cannot say in advance whether or not a set of nonlinear equations will give rise to emergent phenomena; this almost always requires simulating them on a computer. Restated in logical terms, nonlinearity is necessary but not sufficient for emergent behavior. In a world where almost everything is nonlinear, computer simulations are an essential tool for science. Simulation via agent-based modeling provides us with a laboratory to explore the economic world – to boldly go where no one has gone before, and to study the world 'as is' rather than 'as if' it were something else.

When macroeconomists are confronted by nonlinear equations, because they are hard to work with, they often simplify them to make them linear. But because nonlinear dynamics is a prerequisite for endogenous dynamics – that is, motion from within – linearizing the equations makes it impossible for dynamics to emerge. If you want to understand a business cycle, linearizing *guarantees* that the (model) economy won't move by itself. Agent-based models liberate us from these kinds of assumptions.

The COVID pandemic, revisited

In the prologue I described the model we built at the beginning of the COVID pandemic that accurately predicted the hit to the UK economy. As the pandemic wore on, far longer than we imagined it would, it became clearer and clearer how the pandemic, the economy and the actions of governments around the world were intricately interlinked. While we could do the economics well, to make a proper model we needed to team up with epidemiologists. (We did not aspire to predict what the governments would do, only to advise them on what was likely to happen, conditional on their decisions.) To build a better model we teamed up with a group including Alessandro Vespignani, from Northeastern University in Boston. Alessandro is a leading figure in network science and agent-based modeling of pandemics, and his group made some of the most accurate and informative predictions during the pandemic.[9]

In an effort led by Marco Pangallo on our side, and Alberto Aleta from the epidemic side, we did something new: We built a model that simulated both the epidemiology and the economics of the pandemic in tandem (see Figure 6).[10] This model was for the New York metropolitan area and gave predictions for the economic and epidemic impacts on a day-by-day basis. Both models used a synthetic population of 416,442 individuals grouped into 153,457 households.[11] These numbers were picked to match a database of volunteers who consented to have their mobile phones tracked

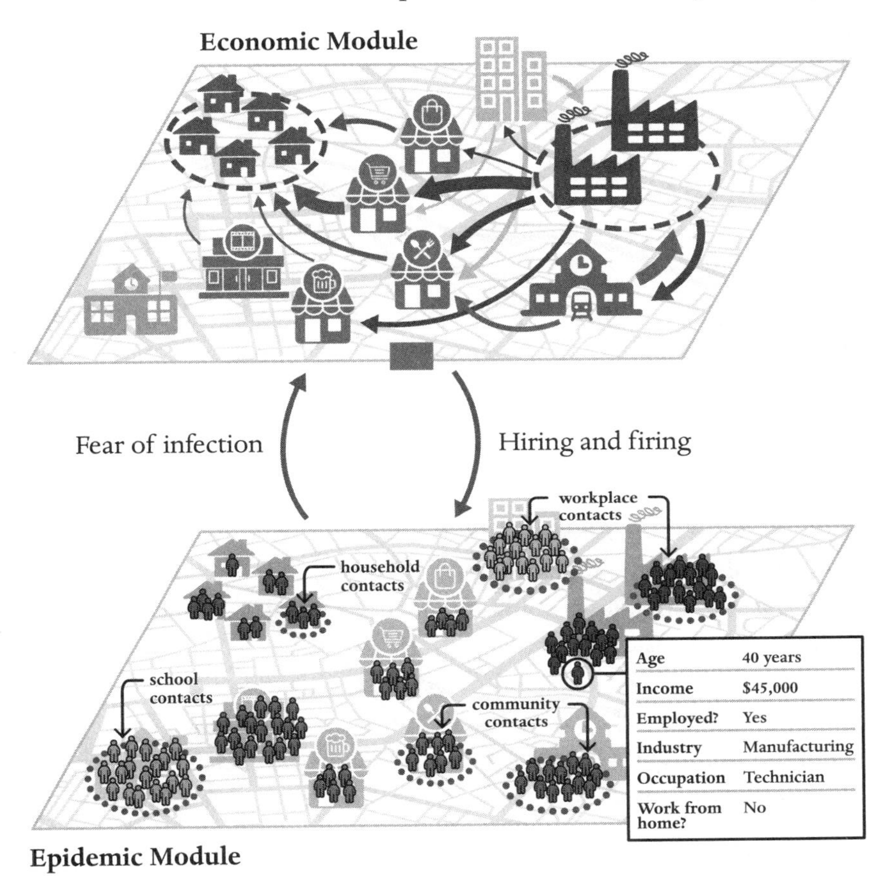

Figure 6. Our model for the epidemic's effect on sickness and the economy.

(though their identities were confidential). As shown in the figure, the synthetic population matched key demographic features of the real population, including age, income, employment, industry sector of employment, occupation and ability to work from home. The movements of the tracked volunteers allowed us to see how much they moved around the city, and therefore how likely they were to infect or be infected by others, and to determine whether they went to work. The models were coupled: If someone becomes unemployed, they cease to go to work and can no longer get infected in the workplace, whereas if someone who is previously unemployed becomes employed their danger of infection increases. The epidemic and economic models send signals to each other every day, which are incorporated into their forecasts for the following day.

Marco took advantage of data on the types of economic activity in each census tract to build a very fine-grained geographic model of the economics of the pandemic. This made it possible to predict how the pandemic would affect the economy geographically, enabling predictions conditional on poverty, race and age. It allowed us to study different policies to search for those that were 'least bad'. We first demonstrated that the model did a good job of predicting the onset of the pandemic in 2020 (though in this case we had the benefit of hindsight, so we were really fitting to the data rather than making a true prediction).

Once we were confident in the model, we investigated a variety of different 'what-if' scenarios. For example, we showed that if the first New York lockdown had only been imposed a couple of weeks earlier, there would have been both far fewer deaths and much less economic suffering. We also showed that the tradeoff between economic suffering and epidemic deaths is similar whether outcomes change due to fear of infection or non-pharmaceutical inventions. In either case, low-income workers experience much higher infections and much worse economic harm.

Our model is innovative because it couples together two agent-based models, one for economics and one for epidemics, to provide useful policy analysis. This could be valuable during future pandemics in guiding policymakers to understand the tradeoff between different sources of harm. It also provides a proof of principle for how agent-based models from different disciplines can work together to provide better policy advice.

A housing bubble from the bottom up

Shortly after the 2008 financial crisis, we built an agent-based model of the Washington DC housing market.[12] There are several reasons why housing markets are an ideal candidate for agent-based modeling. First, as we learned the hard way in the crisis, housing markets are really important to the entire economy. Second, the way houses are bought and sold makes housing markets hard to model using standard theory (more on this in a moment). In contrast, our agent-based model simply mimics the institutional structures for buying and selling houses in the real world and employs rules of thumb that people actually use to make decisions. The model involves a lot of computer programming and data but the way it works will be familiar to anyone who has ever bought a house.

Using detailed data on real-estate transactions and mortgages, and information from the IRS and the US census, our model reproduced all the steps of buying or selling a house. We simulated the decisions of households to rent or buy and their interactions with banks and real-estate agents in taking out loans and buying and selling houses. Although this might sound like an obvious way to do things, it's completely different from any existing mainstream economic model. The result provides useful insight into what drove the housing bubble that peaked in 2006, and how the next one can be avoided.

The housing bubble played a major role in the ensuing financial crisis. In real terms, between 1997 and 2006, US housing prices increased by almost a factor of 3 and then dropped by 30 per cent over the next four years.[13] This had a significant effect on the economy for several reasons.

First, in the United States and many other countries, housing is the biggest repository of household savings. When housing prices drop, everyone who owns a house becomes poorer – so people spend less, which means that companies lower production and unemployment goes up. This means fewer people buy houses, which lowers housing prices further and makes people even poorer, creating a vicious cycle. This happens when any housing bubble bursts.

What made this bubble special, and made the bursting of the bubble particularly painful, is that banks had heavily invested in a recent financial innovation called mortgage-backed securities, which are bundles of mortgages. When the housing bubble burst, the value of mortgage-backed securities dropped substantially. Because they had been viewed as a safe investment, the biggest financial institutions throughout the world held lots of mortgage-backed securities. When their value dropped, this created a global credit crisis; the financial system abruptly switched from easy credit to no credit at all. As a result, many businesses couldn't borrow the money they needed to operate, increasing unemployment, causing less consumption and depressing global economic output. Unfortunately, as noted in the Introduction, when it was consulted about the possible effect of a decline in housing prices in 2006, the Federal Reserve Bank model vastly underestimated how this would affect the economy.

But what caused the housing bubble? Possible explanations include lending practices, high interest rates and demographic shifts. How big a role did each of these play?

My collaborators and I were empowered to address these questions through a grant we received in 2010 from the Institute for New Economic Thinking, which was created with a large donation from George Soros. (This is the global Institute for New Economic Thinking, not the one at Oxford, which was founded two years later.) At the time, I was still at the Santa Fe Institute, where my fellow external-faculty member Rob Axtell was visiting. Rob is one of the pioneers of agent-based modeling (he was doing it long before I was), and he ran the Computational Social Science Program at George Mason University.[14] To complete the roster of senior people on the team for this project, we enlisted the already-mentioned John Geanakoplos, who is, among other things, an expert on mortgage-backed securities and housing markets. Finally, we recruited Peter Howitt, then a professor at Brown and perhaps the only prominent mainstream economist who does agent-based modeling on the side.

Housing markets are a local phenomenon, so we decided to make our model on the scale of a typical American city. The housing bubble did not move in lockstep throughout the United

States – there was a huge bubble in Las Vegas, for example, while the market stayed nearly flat in Dallas. Washington DC was somewhere in the middle, so it seemed like a good choice.

Data collection was a major challenge: We gathered extensive data sets on home prices, lending practices, income, wealth and demographics for the period 1997–2010, including several data sets specific to Washington. (It helped that Rob was based in Washington; his personal connections gave us access to data we would never otherwise have gotten.)

Our goal was to make the most realistic possible simulation of how houses are bought and sold. This meant modeling details like how people got loans from banks and how real-estate agents helped them find the right house. The agents in our model were households, who could be homeowners or renters. We simulated the behavior of 10,000 households in small runs and up to 100,000 in larger runs – fewer than the actual number of households in Washington but enough to provide a good sample. We created artificial households with different levels of income and wealth, so that they were representative of the actual households in the Washington DC area in roughly the right proportions.

The central question that each of our households faced was, 'Should I rent or buy?' Buying might result in a higher monthly payment, but in contrast to renting, some of it goes to pay off the principal (the amount borrowed). The decision of whether to rent or buy is strongly influenced by whether or not you expect prices to rise. If your neighbors own their houses and you don't own yours, and housing prices go up, you'll likely regret not buying. But if you expect prices to fall, then buying may not be wise. We built a model of the decision-making of prospective buyers. Each household estimated future house prices based on the recent trend; if house prices were currently going up, the households assumed they would keep going up, at least for a while. The fact that people make house-buying decisions this way is well supported by the data and is an example of what economists call *backward-looking expectations*. At the same time, our households thought about how future prices would affect them. Thus, our model for prospective buyers used a mixture of forward- and backward-looking expectations, based on research supported by data and common sense.

The fact that most people take out a mortgage when they buy a house raises the stakes. This leverages the investment, amplifying gains or losses. If you put 20 per cent down and prices go up by 20 per cent, you double your money, making 100 per cent on your investment, and you're a happy camper. But if prices drop by 20 per cent, you lose your entire investment. The collapse of housing prices in 2007 caught many people by surprise, and the fact that many of them were strongly leveraged left them in dire straits.

Lending policy affects who has access to credit, something we hypothesized had a huge effect on house prices. To test this hypothesis, we simulated the process of taking out a housing loan. When our simulated households decided to buy a house, unless they were rich enough to buy the house outright, they went to a simulated bank to ask for a loan. The bank would consider each loan application, consider the household's income and wealth, and accept or reject the application accordingly. The simulated bank offered several kinds of loans, ranging from simple, old-fashioned fixed-rate thirty-year mortgages to complicated, newfangled loans with variable rates and balloon payments. In the build-up to the actual crisis, lending policy became more liberal: It was easier to get a loan and the terms were more generous, often with higher leverage and a higher proportion of complicated loans. We were able to acquire a detailed data set with a record of almost all mortgages in Washington, so we could adjust our bank algorithm on a year-by-year basis to match the types of mortgages that were actually issued. We could also run counterfactual scenarios, for example holding the types of loans constant. Similarly, the interest rates at which loans are given could be adjusted to match historical rates, or we could make up hypothetical interest rates to assess their importance in driving the bubble.

Demography also plays an important role in determining house prices. If more people move to a city than move out of it, they will increase demand and drive prices up. We were able to acquire data based on tax returns that told us how many people in a given age and income range entered or left Washington each year.[15] In the course of our simulation, for each year, we could introduce new households (who needed to buy or rent new houses) or put new houses on the market when they moved away from Washington. We also knew how many people put their houses up for sale and remained within Washington.

At the beginning of the simulation, we assigned households diverse incomes and wealth to match the real data in 1997. Then, for each household, we simulated subsequent savings and consumption behavior using the best economic models available.[16] Most households' incomes tended to slowly increase, but unlucky households could become unemployed and default on their mortgages.

Standard models in economics assume that supply equals demand. This is a reasonable approximation for the stock market on timescales of more than a day, but it is a terrible approximation for housing markets. The supply of houses available for sale at any point in time can be very far from the demand, so that at any given moment there can be a substantial mismatch between the number of houses on the market and the number of people looking for a house. As a result, housing prices often change sluggishly in response to changes in supply and demand, and the inventory of unsold houses varies significantly, based on the state of the housing market.

This meant we had to be more realistic in the way we set housing prices. Once again, we closely mimicked the way this actually happens. We were fortunate to obtain data from the association of real-estate agencies in the Washington area that recorded all the quoted asking prices for every house sold through an agency, as well as the final sale price. We analyzed this data and determined that housing prices are set like this: When a house is put on the market, the seller looks for recent sales of comparable houses and marks his or her own house up by a small amount, typically about 5 per cent. Then if the house doesn't sell after a month, the seller marks it down by about 10 per cent, and continues marking it down until either the house sells or is taken off the market. (This will be familiar to anyone who has bought or sold a house, but remarkably, no other economic models set prices this way.)

We also modeled the rental market. This meant we had to model landlords, who buy houses to rent to others. This part of our model was less accurate, because we couldn't get good data on rental prices.

We simulated the model on a monthly timescale. We began by initializing the model to match conditions at the beginning of 1997. Each month, the model would run through a series of steps like this:

1. Demography is updated, so that new households enter and leave the area to match the historical record.
2. The banks survey all their existing loans and make foreclosure decisions.
3. Households are updated. This requires a long list of possible actions and decisions for each household, including: receive income; consume; make a rental or mortgage payment; decide whether or not to buy a new house; decide whether or not to sell, list or delist a house that is already on the market; apply for a loan or refinance an existing loan.
4. The banks consider loan applications and offer terms for mortgages.
5. Houses and buyers enter the market, and houses are bought and sold.

Models always have parameters, which are like knobs that need to be adjusted to specify exactly what the model is. Some of the parameters are easy to pin down, like a tax rate, which you simply look up. Other parameters may not be so straightforward to set; we call these *free parameters.* Our model had quite a few of them. An example is the number of years households use for guessing trends in housing prices: Do people think a rise in house prices last year implies a rise next year? Or do they consider the last four years? Another parameter determines how much the household weights the need for housing relative to other forms of consumption. We could adjust the parameters to closely match the data. However, this didn't necessarily show that the model was good – with enough free parameters, you can fit almost anything to historical data – and we wanted to be sure we were on solid ground.

To address this problem, we set the parameters for each part of the model separately, using microdata. For example, we adjusted the parameters for how the households set their housing budgets by looking at census data (which recorded how much individual households actually spent), and the parameters for their strategies for selling houses by looking at real-estate quotes, and so on. This was a lot of work – even more than constructing the original model. Then we put all the modules together and pushed Run. We started the model in 1997 and let it run without any intervention until 2010. The last value in the simulation was, in a sense, a thirteen-year-ahead prediction.[17] To test the validity of our model, we then compared our simulation's prediction to what actually happened in the Washington housing market during those thirteen years. The results are shown in Figure 7.

The plots shown in Figure 7 consider nine different features of the housing market. The match is by no means perfect; for example, the simulated bubble peaks before the real bubble and prices go a little higher. The model doesn't fit every feature perfectly, but most of the time it's in the ballpark. It is important to look at the scales on the plots – for home ownership rates, for example, we predicted rates around 60 per cent, whereas in reality they varied between 65 and 70 per cent. So we weren't doing badly given that we were missing the data needed to calibrate some of the features, such as the rental market.

In addition to trying to match reality, we also did some counterfactual experiments. When we held interest rates constant at the initial (high) value, for example, the bubble in our simulation was damped. This was expected; low interest rates reduce monthly payments and so create more demand for housing. We also experimented with holding lending policy constant, with banks continuing to issue the same type of loans they were making in 1997, which were

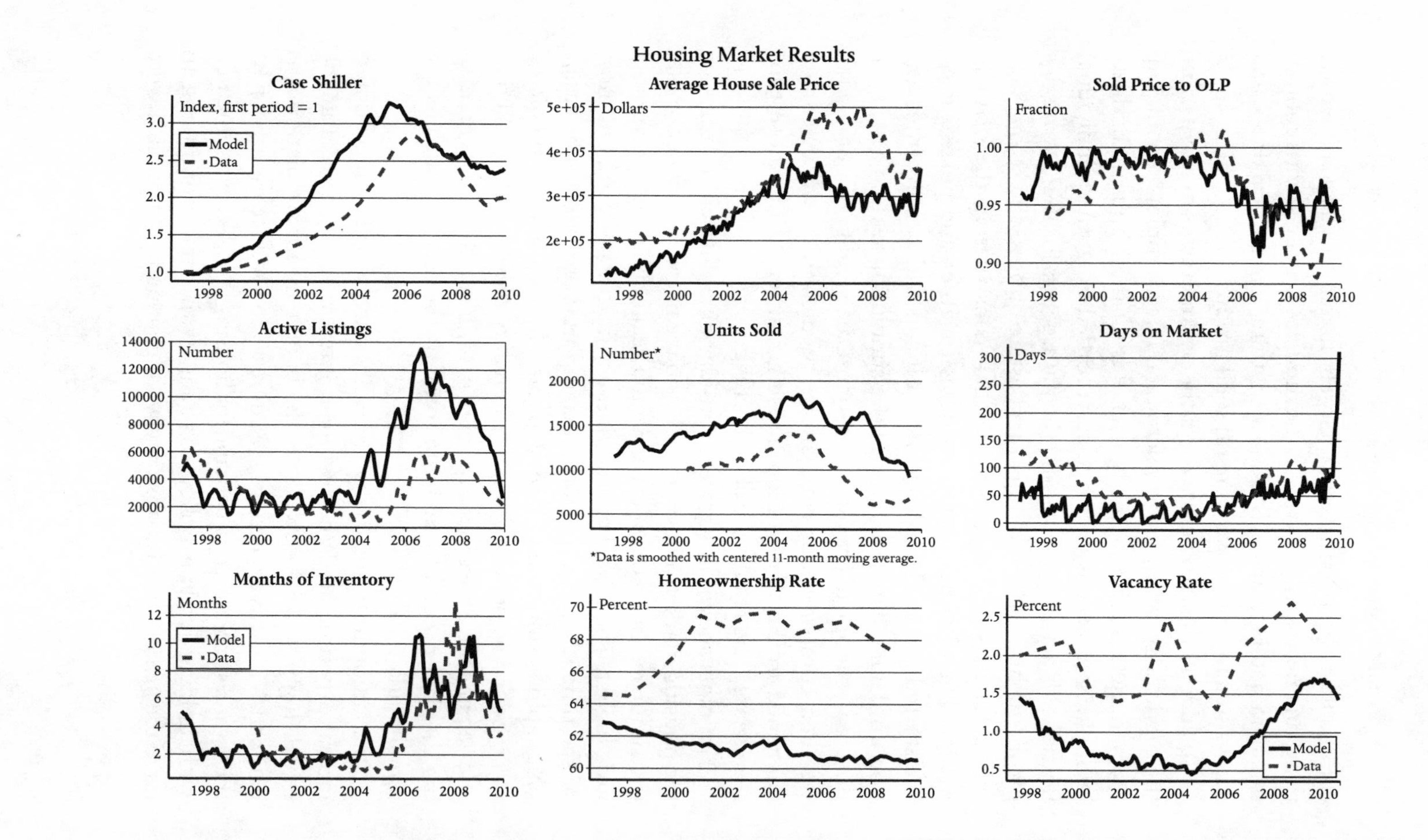
Housing Market Results
Case Shiller
Index, first period = 1
Model
Data
Average House Sale Price
Dollars
Sold Price to OLP
Fraction
Active Listings
Number
Units Sold
Number*
*Data is smoothed with centered 11-month moving average.
Days on Market
Days
Months of Inventory
Months
Model
Data
Homeownership Rate
Percent
Vacancy Rate
Percent
Model
Data

mostly fixed-rate with down payments of 20 per cent or more. The result was that the housing bubble almost disappeared, which confirmed our hypothesis that shifts in lending policy played a bigger role in causing the housing bubble than other factors, such as interest rates. The important role of lending policy is understood now, but it was by no means well established at the time we developed our model.

At this point in the project, we ran out of funds and had to quit. However, this work was continued in a collaboration between my group at Oxford and the Bank of England. We constructed a simpler model based on the same principles, making heavy use of what was learned from modeling the Washington DC market. The Bank of England model was calibrated on British data and used to study lending policies in the UK.[19] Housing prices in the London area had been climbing for two decades, and in 2016 they were extremely high and still rising. With a few caveats, our model was able to reproduce this behavior.

To keep housing prices in the London area from rising even further, the Bank of England was contemplating restricting lending so that 85 per cent of bank loans would have a loan-to-income ratio less than 3.5. In other words, they proposed that, in most cases, to get a loan for £350,000, the borrower would have to have an annual income of at least £100,000. The goal was to dampen the bubble (though the Bank of England forbade us from using the word *bubble*), but not so strongly that the market crashed. We tested this policy in our model by simulating a bubble and then suddenly implementing the proposed policy as prices were climbing. When we did this, prices stopped climbing but didn't crash, which was the

Figure 7. *Housing market simulation for Washington DC.* We compare predictions to data for nine different features of the housing market. Solid lines are model outputs, dashed lines are data. The Case-Shiller index is an indicator of how much housing prices have changed; '1' indicates not at all, '2' indicates that prices have doubled, etc.[17] 'OLP' is 'Original Listing Price'. Note the scale of the plots; for example, it looks like homeownership rates are substantially wrong, but the difference is only about 5 percent.

desired behavior. Our results were presented to the UK Financial Policy Committee and shortly thereafter they implemented the policy. (We were not privy to the committee's deliberations and don't know whether our model influenced their decision.) After the policy was implemented, the bubble flattened out, essentially as our model predicted it would.

This work shows the potential of large-scale modeling of the housing market. With today's computing power, it is possible to simulate markets with hundreds of millions of houses. There exists detailed data about housing transactions and mortgages for the entire United States, which could be used to make a realistic and quantitatively accurate model. Cities like Washington or London can be modeled at one-to-one scale, including the characteristics of each house in its geographic location, to understand how local factors like transportation networks and neighborhoods affect housing prices.

Models like this are now being pursued by at least seven different central banks and could help prevent future housing bubbles.[20] The Bank of Hungary, for example, has an agent-based model that maps the Hungarian housing market at one-to-one scale – each person and each house is represented. They are using this model to predict how government policy levers will affect housing prices and formulate policies to reduce defaults and damp bubbles. Such models are also potentially very helpful to city planners who want to implement policies for affordable housing.

PART II:

Standard Economics ⇔ Complexity Economics

We have so far focused on introducing complexity economics, providing some motivation and a few examples. It may seem that complexity economics is just common sense – it follows a literal, bottom-up approach to modeling the economy that is used in many other fields of science. If you are not an economist, you may be thinking, 'What's the big deal? This seems like the obvious way to do things.' If you are an economist, however, you will recognize that complexity economics is heresy – it violates principles that have been the foundation of the field of economics for more than a century. To put things in perspective, we will now briefly discuss how standard economic theory works.

5.

A Lightning Summary of Standard Economics

To understand why complexity economics is revolutionary, it needs to be compared to standard economic theory, which is built on a set of very different assumptions that developed over the last two centuries. The core concepts were formed before computers were widely used, and when empirical data from the social sciences about human behavior was still scant. To cope with these limitations, economists made bold assumptions. They took a top-down approach, developing a mathematical theory about how people make decisions and how this affects the economy. While economics has evolved substantially over the last fifty years to include such new developments as behavioral economics, the central ideas remain much the same. There is not room here to provide a proper exposition of standard economic theory, but I will present a succinct summary of the main ideas.

Foundational assumptions of standard economic theory

I think of the key assumptions of standard economic theory as three pillars: *utility maximization*, *equilibrium* and *agent beliefs*. These three ideas are taught in all economics textbooks and form the basis for all mainstream theory.

The central pillar is *utility maximization*. Utility is like a scorecard: Given a set of choices, utility measures the relative desirability of each one, so that you can decide which you prefer. Mainstream economists assume that any choice can be given a utility score and that we choose the best option. The idea of utility grew out of the mid-nineteenth-century school of thought called Utilitarianism, which posited that we all seek to be happy, at least in some sense, and that happiness can be measured.[1] The notion of utility used in contemporary economics has evolved since then. It no longer represents happiness, but is instead simply a mathematical device for expressing agents' preferences. Each agent is assigned a utility function that represents the relative desirability of her possible choices. Economic theories are built by assuming that agents make choices that maximize their expected utility – that is, the utility they will get on average given their beliefs about the future.

To understand how this works, imagine you're about to go on a picnic, and you're in the supermarket wondering whether to buy a disposable umbrella or a fancy dessert, which both cost $10. To make a good choice you need to compare your loss of utility from getting wet against your gain in utility from eating the fancy dessert. To do this you need to understand your preferences. You might think, 'If it rains, I would pay $100 to stay dry.' To rephrase this in economics jargon, let's say your baseline utility for being dry is 0, your (dis)utility for getting wet is −100, and your utility from eating the fancy dessert is 10.

To make your decision, for each choice and each possible outcome, you multiply the utility for the outcome by its probability in order to find the decision that results in higher utility. There are four possible outcomes:

- *You buy the umbrella and it rains*. No dessert but you stay dry whether or not it rains, so you get your baseline *utility* = 0.
- *You buy the umbrella and it doesn't rain*. No dessert and you stay dry, so this is the same as above: Your *utility* = 0.
- *You buy the dessert and it rains*. The yummy dessert gives you a utility of 10 but you experience a disutility of −100 from getting wet, so overall your *utility* = −90.
- *You buy the dessert and it doesn't rain*. The yummy desert gives you a utility of 10 and you stay dry, so your *utility* = 10.

If you buy the umbrella then your expected utility is zero whether or not it rains. But your expected utility if you don't buy the umbrella depends on how likely you think it is to rain. If you think that there is a 20 per cent chance of rain then your expected utility is $-90 \times 0.2 + 10 \times (1-0.2) = -10$. Zero is larger than -10, so you buy the umbrella and forgo the dessert. (But if you think the probability of rain is less than 10 per cent you will instead buy the dessert.)

Standard economic models assume that agents go through calculations like this to determine their expected utility for each possible decision, and choose the decisions with the highest utility. In a macroeconomic model an agent might be a firm, whose utility is based on maximizing profits, or a household, whose utility is based on maximizing consumption. Utility maximization is the starting point for virtually every mainstream economic model. Economists like it because it provides a way to model people's goals and infer their decisions.

The second pillar of mainstream economic theory is *equilibrium*. This idea's long history stems from the nineteenth-century French mathematical economist Léon Walras, who developed a theory for price formation based on the way stocks used to be traded in the Paris Bourse. In those days an auctioneer would propose a price for a stock and poll the traders in the room to see how many shares were offered to buy or sell at that price. If there was excess demand the auctioneer would raise the price, and if there was excess supply he would lower it. The auctioneer kept polling the traders until demand matched the supply, at which point he would stop and transactions would take place at the final price. By analogy to physics, Walras called the state where supply matches demand *equilibrium*. The concept of equilibrium was independently developed by others, including the great British economist Alfred Marshall, and is a centerpiece of economic theory.

Marshall put the first two pillars together by showing how equilibrium prices and quantities can be derived from utility maximization, building on earlier work by the British logician William Stanley Jevons. Marshall assumed that the price an agent is willing to pay for a good (like an umbrella) is determined by her utility score for that good. He showed how to use this to compute supply-and-demand curves and find the equilibrium where supply meets demand, making it possible to compute the price of the good and the quantity traded. Because of its convenience and power, the equilibrium assumption soon became standard in economic theory. By the middle of the twentieth century, economists like Paul Samuelson, Kenneth Arrow and Robert Solow (and many others) developed Marshall's approach and applied it to problems such as international trade, market regulation and economic growth.

However, something important was still missing. Up until that point, economic models had assumed a world of perfect certainty. Most models were either set in the present, where the utility of each choice is known, or they assumed a deterministic world where the future was perfectly known by every agent. But in the real world, the future is unknown. How could economists construct models with agents who make decisions about an unknown future?

This caused more careful thinking about how agents form their *beliefs* about the future. Each agent in an economic model needs a model of their own to evaluate the likelihood of future events. This allows them to calculate the expected utility associated with each possible decision by weighting the utility of each choice according to its likelihood, as we did in the umbrella example. The modeling of agents' beliefs about the future, which is used to assign probabilities to events, constitutes the third pillar.

The umbrella example is simple because there is just one agent. But in more interesting settings agents need to take other agents into account. A store selling umbrellas can make higher profits by understanding its customers: If the patrons of the store think rain is likely, they may be able to charge more for umbrellas, but if rain is unlikely, they may need to charge less. To set prices correctly this way, they need to understand how their customers make decisions. This requires a model for agents' beliefs that takes into account agents' abilities to form models for their own beliefs and the beliefs of others.

The most widely used model of beliefs is called *rational expectations*, and was introduced in 1961 by American economist John Muth.[2] Under rational expectations, agents use the information available to them to make the best decisions possible based on expectations that are consistent with their underlying model of the world. This is the most common way economists make models for agents who need to take each other into account.

A study of students' decisions in buying textbooks provides a simple illustration.[3] Publishers like to issue new editions of a textbook frequently, forcing students to buy the new edition. But students have a choice to buy a textbook new or used, or not buy it at all. Unsurprisingly, when publishers issue new editions too frequently, students become reluctant to buy new textbooks because they might go out of date before they can be resold. Publishers recognize that students behave this way, and adjust the frequency of new editions accordingly to maximize their profits. This is a good example of a case where the hypothesis of rational-expectations equilibrium succeeds: Students reason about the behavior of publishers and publishers reason about the behavior of students, and they settle into an equilibrium where their expectations are consistent with each other.

The evolution of contemporary macroeconomics

The rational-expectations model of decision-making caught on quickly in microeconomics, the branch of economics that studies how individual agents make decisions, but macroeconomists were slower to adopt it. Through the 1960s and into the 1970s, models in macroeconomics were still based on statistical models of aggregate behavior (econometrics). Rather than trying to understand causal mechanisms by reasoning from first principles, econometrics simply looks to past statistical data to construct models.

The conceptual framework that dominated macroeconomics at the time was based on the ideas of the English economist John Maynard Keynes, who emphasized the importance of the feedback loop between unemployment and production.[4] If many people are thrown out of work, he reasoned, suppliers have fewer consumers. Therefore suppliers have to decrease production, which means firing more workers, which means decreasing production even more, and so on. If unemployment increases, for any reason, the economy can get stuck in an unfavorable self-reinforcing equilibrium in which production is low and unemployment is high. Keynes's followers created econometric models inspired by his ideas. These models were not based on utility maximization and the reasoning of individual agents; rather, they incorporated relationships between factors like unemployment, GDP and interest rates, based on what had happened in the past.

Keynesian-inspired econometric models were challenged by the crisis of high unemployment in the 1970s. In 1974, US unemployment jumped from its previous level of about 4 per cent to almost 8 per cent. This put policymakers under pressure to respond. As the story goes, they tried to exploit something called the *Phillips curve* to bring down joblessness.[5] The Phillips curve refers to the hypothesis that unemployment and inflation are negatively correlated, which up until then had always been the case. If unemployment becomes low, this should push wages higher, which in turn causes inflation. But perhaps causality also runs in the other direction? If prices go up, would manufacturers see this as an optimistic sign and boost production, thus hiring more workers? Policymakers gave it a try, intentionally inflating the money supply in order to spark inflation and drive unemployment down. But this policy failed to lower unemployment, instead causing a nasty combination of joblessness and inflation that was dubbed *stagflation*.

As crises tend to do, this spurred a big shift in economic thinking. Prominent economists such as Robert Lucas of Carnegie Mellon argued that this approach had failed because it neglected the consequences of human intelligence. In his words, employers and workers alike 'saw through the veil of inflation': If the price of everything doubles, and wages double, then nothing has really changed. People realized that what the policymakers had done was artificial, and it just made a bad situation worse.

While the Keynesian econometrics models might fit historical data, such models ignored the possibility that decision-making might change when circumstances did. Lucas and his allies wanted macroeconomists to take human reasoning into account, and proposed to do this using rational expectations. The Lucas critique revolutionized macroeconomics. From the mid-1970s onward, rational expectations took over macroeconomics as well as microeconomics, leading to a new family of models called *Dynamic Stochastic General Equilibrium* (DSGE) models. (*Stochastic* is just a synonym for *random*.) In a standard DSGE model, households and companies alike use rational expectations to make decisions that maximize their utility. In contrast to their predecessors, which were based on statistical relationships of aggregate measures like GDP, these models incorporated the reasoning of households and firms, and for this reason are said to be *micro-founded*.

A simple standard macro-model

To understand how a standard micro-founded DSGE model works, let's look at an early (and simple) example that grew out of an influential model for economic growth introduced in 1956 by Nobel prize-winning economist Robert Solow.[6] (We're going to need this later when we compare it to a complexity-economics version.) In Solow's model, a household works for a company that produces a single good. The firm and household represent all the firms and households in the world (or at least in a given country), and the good represents all goods and services. This is what is called a *representative-agent model*, in which a single agent serves as a proxy for many different agents. There are obvious problems with this, but the goal of the Solow model was to provide insight by making things simple enough to be easy to understand and solve.

In the Solow model, the firm pays the household for the labor and capital it needs to produce the good. A key parameter, which you can think of as an adjustable knob, is the savings rate – the fraction of its income that the household saves. The model also has a parameter for population growth (imagine packing the household with many millions of identical occupants). The Solow model predicts that countries that save more become relatively prosperous and countries with rapid population growth become relatively poorer, and shows that the only way the economy can grow in the long run is by making technological progress. (In this model, technological progress corresponds to producing more of the single good with less capital and labor.)

In the early 1960s, David Cass and Tjalling Koopmans each independently modified the Solow model to turn it into a micro-founded DSGE model. They did this by giving the household a utility function and assuming that it is a rational agent whose goal is to adjust its savings rate in order to maximize its utility.[7] The household's utility at any point in time depends on a combination of its current consumption and its expected future consumption. Present consumption is considered more valuable than consumption in the future, which is called *discounting*. The use of discounted utility was inspired by a 1928 paper by the British mathematician Frank Ramsey, so the resulting model is called the Ramsey-Cass-Koopmans model, or simply the RCK model.[8]

In the original Solow model the savings rate is constant, but in the RCK model the household can adjust it through time. To maximize its utility, the household needs to strike the right balance between savings and consumption. On one hand, it must save to accumulate capital, but on the other hand, it needs to consume to get utility. In the extreme case, if the household doesn't consume anything, the business it owns loses its only customer; as a result, the household makes no profits and has nothing to save. There is an optimal savings rate in the middle, which is the equilibrium choice that maximizes the discounted utility.

If we now add shocks – for example, by letting the discount rate fluctuate randomly – then the RCK model becomes a DSGE model. Each external shock knocks the system away from equilibrium, but the rational agent immediately adjusts its savings rate to compensate and get back on the best path. Because there are ongoing shocks, the economy is always changing; but if we turn off the shocks, the system will eventually come to a state of rest – like a rocking chair.

Friction and other enhancements

Despite early enthusiasm for micro-founded models, during the 1980s it became clear that they had problems matching reality. Given their simplicity, this was not surprising, even to their advocates. While they captured some features of the economy, there were many others that eluded them. For example, in the early models, as the situation changes, wages change freely to maximize everyone's utility. In the real world, however, workers get upset if their wages are lowered, so a company is reluctant to lower them. This also makes them hesitant to raise them, fearing that it might need to lower them again later. Wages play an essential role in inflation, so the mismatch between the models and reality was a serious problem.

In 1983, the Argentine-American economist Guillermo Calvo found a way to fix this.[9] He modified a standard micro-founded model so that wages no longer changed in response to each fluctuation in supply and demand. Instead, they normally remained stuck, but once in a while, at random, they suddenly adjusted to the proper equilibrium level where wages were set so that the demand for labor matched the supply of labor. His hypothesis that wages are 'sticky' is just one example of what economists call *frictions*. In the presence of friction, agents still maximize utility using rational expectations, but real-world constraints prevent them from doing what they would otherwise like to do. Economists have incorporated frictions in models to make the behavior of agents more like that of real people, who act 'as if' they are not fully rational. To return to *Star Trek*, using frictions is like putting Mr Spock in chains: He's still rational, but many solutions are now off limits, so he has to work around them to find the best option.

The use of frictions was initially controversial, even within the mainstream. One of the original motives for rational expectations was to avoid ad-hoc assumptions about behavior, but frictions are ad hoc by their very nature. Calvo's critics ridiculed his sticky-wage assumption as depending on a 'Calvo fairy' to sprinkle pixie dust on frozen wages to release them. Nonetheless, frictions have now become an accepted part of the macroeconomics toolkit. Choosing the right set of frictions and solving the resulting models in order to better fit the data is a main line of research in theoretical macroeconomics.

Modern micro-founded macro-models used by central banks are more complicated than the simple RCK model. Such models are now used by virtually all central banks in the world.[10] Their forecasts are at best a little better than those of simple statistical models, but unlike statistical models, they can be used to test policies with 'what-if' experiments. (At least in principle: As noted earlier, if their forecasts are poor, their reliability for policy testing is questionable.)

Models at the frontier of macroeconomic research incorporate household diversity, for example, by allowing two types of households. The rich household makes savings decisions based on rational expectations and the poor household lives hand-to-mouth, meaning that they do not save anything. In what are called Heterogeneous Agent New Keynesian (HANK) models there is a continuous distribution of different households, ranging from rich to poor. Nonetheless, like the Solow model, these models still have only one representative industry. At the other extreme, there are models with many industries, but only a single representative household.[11] There is another class of models, called Computable General Equilibrium models, that allow multiple industries and multiple households, but they have no dynamics (meaning that the models cannot study the path to a stationary equilibrium – they jump to the static equilibrium in one step). As we'll see in the next chapter, these limitations stem from technical difficulties in building these kind of models (which agent-based models avoid).

All micro-founded macro-models, even those at the frontier of research, are based on rational expectations with frictions, and incorporate little or no information about the non-rational behavior of real people. Changing this is another important frontier of contemporary research.

The behavioral revolution

As rational-expectations theory took over macroeconomics in the 1970s, dissenting voices began to speak out and gather evidence against it. Psychologists Daniel Kahneman and Amos Tversky documented deviations from rationality in decision-making, forming a close partnership with behavioral economist Richard Thaler. There is now a long list of common deviations from rationality, all backed by experimental evidence. People tend to be overconfident, particularly men; for example, students do not follow rational expectations when forecasting their grades – they consistently overestimate their final grade, even though it should be fairly predictable based on factors such as past performance, ability and effort level.[12] People pay inappropriate attention to sunk costs: If you're playing poker, for example, you should ignore how much you've already put in the pot and worry only about whether your next bet is a good bet. These are just two examples of many.

One of the big innovations of behavioral macroeconomics was the use of experiments, pioneered by Vernon Smith.[13] Up until his work, it had been assumed that economies could not be studied in the laboratory. (The analogy was made to astronomy: We can observe the stars, but we can't do experiments with them.) Smith put his students in controlled environments, such as asking them to play a game that was like a real market. He then asked them to make economically relevant decisions (such as whether to sell an asset and at what price) and rewarded them based on their performance. His experiments made it clear that markets converge to equilibrium in some situations but not in others.[14] Experimental economics has been used to indicate when and how the economic behavior of *Homo sapiens* departs from that of *Homo economicus*.

Not surprisingly, economics experiments were initially highly controversial. Were the rewards large enough to be meaningful? Would people really behave in the real world as they did in the laboratory? Despite the initial resistance, experimental economics and behavioral economics slowly entered the mainstream. Kahneman and Smith shared the Nobel Prize in economics in 2002, and Thaler was awarded it in 2017.

Behavioral economics has now made the serious flaws of rational-expectations theory quite clear, but so far it has proved difficult to incorporate behavioral alternatives into the workhorse macroeconomic models used by central banks and treasury departments. Achieving this is a central quest in contemporary economics, but it remains largely unsuccessful. Economists often say that 'it takes a model to beat a model', and there is still no widely accepted model that replaces rational expectations with behavioral economics.

In the heyday of rational expectations, the three pillars of mainstream economics fitted together nicely. To recap, in those days there was a straightforward procedure to build an economic model:

1. Give each agent a *utility* function.
2. Use *rational expectations* – i.e., assume each agent's beliefs are correct and find the set of decisions that maximizes each agent's utility.
3. Solve for the *equilibrium* values of the relevant economic variables (like prices and quantities traded).

Responding to the challenge posed by behavioral economics, some economists have proposed replacements for step 2. One popular alternative is to let some agents be rational while others are not. In recent years, economists have begun trying to incorporate the findings of behavioral economists more directly, by assuming that people make decisions based on flawed models of the world. For example, Xavier Gabaix has developed a macroeconomic model with myopic households who fail to look very far into the future.[15] Similarly, Nicola Gennaioli and Andrei Shleifer have created a 'diagnostic expectations' model, incorporating departures from rational expectations by assuming that people overreact to news.[16] These approaches replace rational expectations with a different model of beliefs, but they still assume that agents maximize utility, or something like utility.

This brings us to the present state of mainstream economics. As I write, in 2023, two of the three pillars that form the foundation of standard economic theory remain firmly in place, much as they were sixty years ago. Virtually all mainstream economic theories continue to subscribe to utility maximization (or some close cousin) and equilibrium. Most mainstream theorists still assume rational expectations, though some explore alternatives to the third pillar, agent beliefs. In contrast, as discussed in more detail in the next chapter, complexity economics modifies or replaces all three pillars.

Value-free modeling?

Economic models are often viewed with suspicion because they reflect the values of their creators. The austerity debate, which we discussed in Chapter 1, is just one of many examples of prominent economists reaching opposite conclusions – conclusions that reflect their values more than any underlying science. In an essay titled 'Economics after Neoliberalism', Suresh Naidu, Dani Rodrik and Gabriel Zucman unintentionally make this point:

> Back in 1975, economist Carlos F. Diaz-Alejandro wrote, 'by now any bright graduate student, by choosing his assumptions [. . .] carefully, can produce a consistent model yielding just about any policy recommendation he favored at the start.' Economics has become even richer in the intervening four decades.[17]

This kind of 'richness' concerns me. If left-wing and right-wing economists can build models for the same thing 'yielding just about any policy recommendation [they] favor at the start', the value of economics as a reliable roadmap for all to share is seriously limited. It's not very useful to have one version of economic science for those on the right and another for those on the left. As almost all economists will agree, we should strive to be as objective and fact-driven as possible. This is one of the biggest reasons theories need to be empirically validated.

In fact, Naidu and his co-authors were trying to make a different point – one that I agree with. They were explaining how economics has evolved beyond the simple-minded misconceptions of neoliberalism. The underpinnings of the Chicago school of economics, led by Milton Friedman, used idealizations – such as perfectly functioning markets and rational expectations – that led to conclusions supporting the school's libertarian leanings. But once economic models began to incorporate more realistic assumptions, the results became more nuanced and conditional, and most of the earlier results were shown to be wrong. This was an important step in the right direction.

It will always be possible for researchers to fiddle with their models until they get preconceived, desired results. But since complexity-economics models can be founded directly on data from behavioral experiments, and can match the structure of the economy more literally, this becomes harder to do. In a complexity-economics model it is more obvious if someone is cooking the books to get the answer they want. So my hope is that complexity economics can provide a more objective framework; time will tell whether this hope is realized.

Uncertainty vs risk

[A]s we know, there are known knowns; there are things we know we know. We also know there are known unknowns; that is to say we know there are some things we do not know. But there are also unknown unknowns – the ones we don't know we don't know. [They] tend to be the difficult ones.

Donald Rumsfeld (2002)[18]

In 1921, the economist Frank Knight, who would go on to be a co-founder of the Chicago school but was at that time at the University of Iowa, was the first to make a clear distinction between risk and uncertainty.[19] By *risk* he meant situations where we know the set of possible future events and their probabilities. In contrast, he used *uncertainty* to refer to situations where we don't know the probabilities of future events. Under what he called *true uncertainty*, we may not even be able to imagine all possible future events.

The standard workhorse macroeconomic models that are currently used by central banks and treasury departments assume that we live in a world of risk rather than one of uncertainty. Even if Zeus is throwing thunderbolts at the world, we still know the possible events that can occur as a result and the likelihood with which they will occur. These models assume that agents make decisions as described earlier in this chapter, by listing all possible future events, estimating their probabilities, calculating the utility of each possible event, multiplying each utility by its probability and adding them up to get the expected utility. The agents then select the choice that maximizes their expected utility.

There is a large literature trying to deal with uncertainty in economic modeling, but this is so far not sufficiently developed to be incorporated into the workhorse models of macroeconomics. As discussed in the next chapter, uncertainty is implicitly built into complexity economics – a feature that makes it an attractive alternative for addressing big problems like climate change where we cannot know all possible events, much less assign them accurate probabilities.

6.

Modeling an Uncertain and Complicated Economy

[T]he task is to replace the global rationality of economic man with a kind of rational behavior that is compatible with the access to information and the computational capacities that are actually possessed by organisms, including man, in the kinds of environments in which such organisms exist.

Herbert Simon (1955)[1]

Even as the theory of rational expectations was being developed in the mid-1950s, its limitations were being criticized. One of the most prominent critics was Herbert Simon, who is perhaps *the* pioneer of complexity economics, even if that term didn't exist in his day.[2]

Simon held that in real situations people often make complicated decisions with imperfect knowledge, noting that 'the capacity of the human mind for formulating and solving complex problems is very small compared with the size of the problems whose solution is required for objectively rational behavior in the real world – or even for a reasonable approximation to such objective rationality'.[3] Simon argued that when we are confronted with a hard problem, the cognitive limitations of the mind make rationality moot. In his view, we seek 'good enough' choices rather than optimal ones. He called this *bounded rationality*. His views on the subject have been confirmed by psychologists and behavioral economists alike. Yet despite his 1978 Nobel Prize, Simon is not a central figure in mainstream economics – perhaps because his approach would have required abandoning the standard theoretical assumptions that captured the mainstream. Fortunately, Simon's research program is far easier to implement now than it was in the 1950s, because we now have a much better understanding of psychology, as well as more data and far better computer power.

Heuristics

Real people don't solve day-to-day problems by doing complicated math calculations. As a simple analogy, let's consider two ways of playing baseball. Imagine Mr Spock in center field. Let's contrast his rational approach to that of a human baseball player.

When a fly ball is hit, Spock knows that his team's utility will be maximized if he catches the ball before it touches the ground. To make this happen, he measures the position and velocity of the ball. He then solves Newton's laws to compute the ball's estimated landing point and runs to that point in the field.

This is how we predicted the game of roulette using a computer in Chapter 1, but it isn't the way real center-fielders catch baseballs. A human baseball player uses simple rules of thumb that she may not even be consciously aware of. For example, she waits until the ball is sufficiently high in the air, then fixes the angle between her body and the ball and begins running to maintain this angle as consistently as possible. If she succeeds, she'll automatically be in the right place to catch the ball.

This rule of thumb is called the *gaze heuristic*.[4] Baseball players use it (along with other rules of thumb) to help catch fly balls. Dogs use it to catch Frisbees, and sailors use a similar technique to determine whether they're on a collision course with another boat. Unlike Mr Spock's method, it doesn't require any math calculations and takes complicating factors like wind into account. It doesn't always result in optimal behavior – a player may first back up and then run forward – but it is useful in practice.

The term *heuristic* comes from a Greek word meaning 'to find out or discover'. Heuristics are simple mental processes that humans, animals and organizations use to quickly form judgments, make decisions and find solutions to complex problems. We use heuristics all the time. To decide whether or not to buy an umbrella, trial-and-error provides a simple alternative to rational expectations: If you have recently been drenched, you are more likely to remember to carry your umbrella. Or you might observe that your friends all carry umbrellas and imitate them. Imitation and trial-and-error are examples of common heuristics that work well in many different contexts. In the next chapter we'll revisit the Solow model, and see how an agent-based model that replaces rational expectations with these two simple heuristics leads to a surprising example of emergent behavior.

We've already seen a powerful and widely used heuristic in action, in the way real-estate agents set house prices, initially offering the house at a price slightly higher than the current market. If the house doesn't sell, they mark it down and wait for a buyer, repeating the process until it finally sells. Psychologists call this heuristic *aspiration-level adaptation*. It is commonly used in many other contexts, like selling used cars. Aspiration-level adaptation proved very useful in our agent-based housing model, since it allowed housing prices to roam far from the point where supply equals demand, better matching reality (see Chapter 4). Although aspiration-level adaptation is easy to simulate in a computer, it isn't easy to write down and solve equations for it. As a result, because of their desire for models with a mathematical underpinning, mainstream economists are loath to use it.

A long list of heuristics have been identified by psychologists, with names like *the availability heuristic, the representativeness heuristic, the anchoring heuristic, the affect heuristic, the consistency heuristic* and *the control heuristic* – and many others.[5] Unlike utility maximization using rational expectations, which requires knowledge that most of us don't have and math calculations that most of us can't do, heuristics are quick and simple, operate on limited information and require little or no math. *Heuristics have verisimilitude.* Psychology experiments show that real people rely on them. They're not the only way we solve problems, but they're an essential component.

You might think that because heuristics are simple, they cannot be a good way to predict the future. But in uncertain environments simplicity is a virtue.[6] As psychologist Gerd Gigerenzer and his colleagues have documented, there are many situations where heuristics make better predictions than more complicated methods. This is called the *less-is-more effect*.[7]

Portfolio choice provides a great example. Given a choice between different assets, like stock or bonds, how much of each should you hold in your portfolio? Suppose, just for the sake of argument, that you know their expected future returns, as well as their volatility and their correlations with each other. Harry Markowitz of UC San Diego got the 1990 Nobel Prize in economics for using rational expectations to find an optimal formula for how much of each asset you should buy in order to make the largest possible return on your portfolio (on average) for a given level of risk.[8] This is a simple and elegant result – but there's a problem.

Later, in Part III of the book, I will describe my adventures as a quantitative financial trader, but let me give a little preview that illustrates how a simple heuristic outperforms Markowitz's 'optimal' result. In 1991, I was one of the founders of an enterprise called Prediction Company. In our early days, we developed strategies for trading currencies, and we wanted to decide what fraction of our trading capital to allocate to the model for each currency. I bought a standard textbook on finance written by Eugene Fama and read the chapter on portfolio theory.[9] It presented Markowitz's method as the right way to solve this problem, without any caveats. I tried it and discovered that it worked fine – as long as the expected return and risk for each currency strategy were known perfectly. But when I made a more realistic test, by pretending I was at some point in the past and didn't know what would happen in the future, the results were terrible.

The reason for the poor performance of Markowitz's formula is uncertainty. At Prediction Company, it was impossible for us to know precisely how well the strategy we applied to each currency would actually perform in the future. Based on Markowitz's formula, small variations in the expected risk and return made the allocation to each currency vary wildly. In fact, the trading system always did better if we used the simple heuristic of assigning equal weights to each currency. Later I found out that Markowitz himself knew the same thing: Ironically, when he invested the money from his Nobel Prize, he used a portfolio with equal weights to pick a few of his favorite stocks rather than the formula he received the prize for discovering.[10,11]

This brings up a key point: When testing a model, one should focus on its ability to make predictions about the future, rather than its ability to fit past data.[12] Unfortunately, most studies in economics don't do this. Models that fit past data well but fail to make good predictions on data they have not seen are worthless. Heuristics often perform relatively poorly at fitting data, but they nonetheless frequently make better predictions of the future. The portfolio example is just one of many where heuristics outperform rational expectations in realistic settings. This is because our library of heuristics has developed through a process of cultural evolution to help us deal with uncertainty, which is a pervasive feature of the real world.[13]

Modeling behavior directly

Heuristics are a good example of what I call *modeling behavior directly*. In contrast to utility, which is cumbersome to observe and must be inferred indirectly, behavior is observable and easily measured – in the laboratory or using data collected via surveys, censuses, the internet, or any forum that records people's decisions. Modeling behavior directly means developing rules that encapsulate the way people actually make decisions. Utility theory, in contrast, deduces people's decisions from their preferences. Utility theory often relies on 'as-if' arguments, while modeling behavior directly is grounded on 'as-is' arguments, at least when it is done well.

Of course, heuristics are only part of modeling behavior directly. Even if our ability to reason is finite – we are only boundedly rational – we nonetheless use reasoning to solve many problems, and we need to take this into account. We have a variety of different methods for doing this; we have ways of deciding which heuristics to use in different situations, and we combine heuristics with other (usually simple) modes of reasoning.

As an example, consider how people – as opposed to computers – play complicated games like chess. There are about 10^{40} possible sequences of moves in a chess game; even the most powerful supercomputer can examine only a tiny fraction of such sequences. Human chess players solve this problem using a mixture of experience and intuition. The goal – to checkmate your opponent's king – looms in the distance at the start of the game, which on average takes about 40 moves but can take as many as 200. A human player can look only a few moves ahead – seeing further becomes too difficult because of the explosion of possibilities. The problem must be broken into pieces and addressed one step at a time.

Checkmate usually requires first gaining an advantage over the opponent. This can be a piece advantage or a position advantage, or a combination of the two. To gain a piece advantage, we're guided by heuristics such as 'a queen is better than a rook'; to gain a position advantage, we use heuristics such as 'control the middle of the board'. This narrows things down so that we need consider only the most promising lines of attack – seeing whether we can swap a rook for a queen, for example, or positioning pawns to better control the middle. In this case, heuristics guide our limited powers for logical reasoning so that we use them more effectively. And as we gain experience, we develop intuition that allows us to search more efficiently and to know which heuristics to use when.

Following in Herbert Simon's footsteps, Gerd Gigerenzer and economist Reinhard Selten have proposed that an essential part of problem-solving is our ability to draw on an 'adaptive toolbox' of heuristics.[14] The toolbox is adaptive because the heuristics we deploy depend on the situation. Our intuition helps us guess which heuristics to apply, and we learn by experience, favoring heuristics that have been successful recently and abandoning those that have performed poorly. Gigerenzer calls this way of solving problems *ecological rationality*.[15] The word *ecological* emphasizes that our cognitive strategies exploit the representation and structure of information (heuristics, search, learning, simple reasoning) to make reasonable decisions. Unlike rational expectations, ecological rationality allows us to cope with the fundamental uncertainties we encounter in day-to-day life.

Learning is another important way we solve problems. This has been extensively studied in psychology, computer science and behavioral economics, leading to the development of standard algorithms that characterize how people learn. A good example is *reinforcement learning*, in which actions that have worked well in the past become more likely and actions that have worked poorly become less likely. While reinforcement learning doesn't fully capture how real people behave, it provides a reasonable approximation in many circumstances.

In the early part of the twenty-first century the field of machine learning – which is essentially the application of sophisticated learning algorithms – has exploded. Problems that previously seemed out of reach in artificial intelligence, like facial recognition and speech recognition, are now routinely solved by artificial neural networks (which are just another type of learning algorithm). Complicated neural-network models that have been trained on a vast corpus of human text, like ChatGPT, are writing student term papers and making tasks like computer programming much easier.

Mainstream economists are beginning to use machine learning for statistical data analysis, as an alternative to traditional econometrics. That said, with a few exceptions, using learning algorithms to simulate the behavior of agents in models is the domain of agent-based modeling.[16,17]

Learning algorithms are easy to incorporate into agent-based models. For example, a corporation might be assigned the goal of maximizing profits. The model could then be given a learning algorithm that uses experience over time to arrive at better decisions. This is a bit like utility maximization, except that it doesn't guarantee perfection – the decisions can be far from optimal – and it can be applied even in complicated settings, using information that is plausibly available. The explosive advances in machine learning provide a huge opportunity for agent-based modeling that is still in its early stage. The challenge is not to create models where the agents use the 'smartest' possible learning algorithms, but rather to faithfully mimic the way people actually make decisions.

Modeling behavior directly is still an unfinished research program.[18] So far, there's no one-size-fits-all method that tells us how people will make decisions in any given situation. In the meantime, agent-based modelers have cobbled together their own models of behavior, which incorporate many different elements. Here are some of the ways agent-based modelers incorporate behavior:

- *State the obvious.* There are many circumstances for which there is an obvious decision rule. For example, a class of stock traders called *value investors* (think Warren Buffett) follow the rule 'buy undervalued assets'.
- *Ask experts,* like psychologists and sociologists, who have done lab experiments or field studies or performed surveys to understand how people made decisions in a similar context.
- *Study data where similar decisions have been made.* In our housing model for Washington DC, for example, we used data from the US census on people's housing choices, conditioned on income and wealth.
- *Do experiments in the context of the model.* One can slow down an agent-based model to human speed, unplugging the computer's decision-making algorithms and replacing them with the decisions of real people, and then build a model mimicking the choices of the human decision-makers. This is a promising approach that remains largely unexplored.[19]

- *Use learning algorithms* to train agents to make better decisions through time. This is the closest thing to a one-size-fits-all approach, and is the way agent-based models answer the Lucas critique. Although there are many learning algorithms, they often lead to similar behaviors.

In mainstream economics there is a very clear and explicit formula for modeling behavior, but the procedures used in agent-based modeling to model behavior directly, as the list above shows, are less cut and dried. They require more judgment on the part of the modeler. At the end of the day, what matters is whether the models are able to represent the economic world with sufficient verisimilitude to make good predictions.

Exploring the wilderness of bounded rationality

If instead we take the view that prices themselves may adjust sluggishly, we enter the wilderness of 'disequilibrium economics'.

Chris Sims (1980)[20]

Although they have similar goals, mainstream economics and complexity economics model the world completely differently. To see the difference clearly, let's recap the standard modeling template from Chapter 5, updating it to account for the rejection of rational expectations by behavioral economics:

1. Give each agent a utility function (or some other way of ranking the desirability of different outcomes).
2. Give each agent a model of beliefs (which might be based on rational expectations or on something more behaviorally grounded) and find the set of decisions that selfishly maximize each agent's utility, subject to these beliefs.
3. Solve for the equilibrium values of the relevant economic variables (like prices and quantities traded).

Complexity economics replaces this with agent-based modeling, which follows a completely different template:

1. Assign each agent a method for making decisions. This can involve heuristics, simple reasoning, search, learning, or a mixture of any of these. When learning algorithms are used, goals may need to be assigned to some agents, but these need not be utility functions.
2. Mimic the institutions, market mechanisms and other economic factors that are actually involved, with as much verisimilitude as necessary (remembering that it can be counterproductive to introduce details that don't matter).
3. Simulate the collective interactions of the agents as time unfolds.

Both methods have their pros and cons. In simple settings, it is reasonable to model real agents 'as if' they follow rational expectations. If things change sufficiently slowly, equilibrium is a plausible assumption. Then we can use the standard framework to write down equations and solve them. This has the advantage of being simple, comprehensible and easy to implement.

In contrast, agent-based modeling is more useful in complicated settings where agents do not plausibly understand things well enough to reason 'as if' they follow rational expectations, or where things are changing sufficiently fast that equilibrium is not a plausible assumption. Agent-based modeling more easily captures the rich structure and interactions that play important roles in the real economic world. There is much more latitude in how things can be done, depending on the problem that needs to be solved. A key difference is in step 3. In the standard template, equilibrium is assumed from the outset – there is no consideration of what might happen outside of it. The complexity-economics template makes no such assumption; rather, as the collective dynamics unfold, equilibrium is a possible outcome, which may or may not be attained.

I like to visualize the quest to find better models of human reasoning as exploring a wilderness, like the cartoon in Figure 8. This metaphor is inspired by a warning that is given to young economists who are tempted to venture into complexity economics: *There is a danger of getting lost in the wilderness of bounded rationality.* There is only one way to be rational, but there are an infinite number of possible ways to be boundedly rational. This makes it easy to get lost. Hence, aspiring economists are encouraged to stick to the well-known path offered by the standard framework.

This advice is partly sound. The human mind is complicated, and until we have a good model of how it works, well supported by empirical findings, there is a danger of getting lost by trying out half-baked models. Rational expectations is like a gate leading into the wilderness of bounded rationality, which provides a useful reference point. By keeping this gate carefully in sight, the mainstream hopes to avoid getting lost. It is a poor approximation of reality when the economic setting gets complicated, but it is a useful guide in simple settings. It provides a well-defined point of departure that can be built on to incorporate more realistic assumptions and get closer to the truth.

But this advice is missing something important: Rational expectations is not the only way to enter the wilderness. Complexity economists like to enter through a different gate, on the other side of the wilderness, by assuming instead that people make decisions in very simple ways. Heuristics, for example, require very little reasoning to implement and sit at the opposite end of the intelligence spectrum from rational expectations. As an extreme case, starting with 'zero-intelligence' agents, who literally make decisions at random, as if they were throwing darts at a dartboard, can be very useful. Maria's occupational labour model in Chapter 3 provides a good example: Workers simply apply for jobs at random – their only intelligence is in knowing

Figure 8 (facing page). *The wilderness of bounded rationality.* The human brain is complicated, like a wilderness, and creating a reliable model of the way we make decisions is difficult. To avoid getting lost, it helps to start with simple assumptions and build incrementally more complicated models. Rationality is the conventional starting point used in economics, but heuristics, and even the extreme assumption of zero intelligence (random decision-making), provide useful and much less developed alternative reference points.

ZERO INTELLIGENCE GATE
THE WILDERNESS of BOUNDED RATIONALITY
RATIONALITY
RATIONALITY GATE

which jobs they are suited for. Such zero-intelligence models also provide a well-defined point of departure that can be built on by incrementally adding learning or incorporating more knowledge from behavioral economics.[21] They provide an alternate gate into the wilderness of bounded rationality.

To make good economic models we should use both gates, depending on which is more effective for solving the problem at hand. In Chapter 8 we will even see an example where it was useful to go through both gates at once and meet in the middle.[22] To capture the complex and complicated nature of real economies, we need to thoroughly explore the wilderness of bounded rationality, entering through both gates.

Tractable models for complicated economies

The real economy – the one that we live and work in – is a complicated beast with many interacting moving parts. The agents whose behavior needs to be modeled include firms, households, banks, financial investors and governments. There are many different kinds of firms producing many different goods and services. Poor people behave differently than rich people, and old people behave differently than young people. Financial globalization and trading connect the economies of 195 different nation-states. Building a macroeconomic model that takes all this into account is challenging.

I think there are serious limitations to utility maximization as a foundation for quantitative models, but my opinions on this are controversial, and since this book is not about criticizing mainstream economics, I have reduced them to an endnote.[23] There is, however, one problem with utility maximization that is uncontroversial: In complicated situations, the equations become very hard to solve. To quote two prominent economists, Xavier Gabaix and David Laibson:

> Tractable models are easy to analyze. Models with maximal tractability can be solved with analytic methods – i.e. paper and pencil

> calculations. At the other extreme, minimally tractable models cannot be solved even with a computer, since the necessary computations/simulations would take too long. For instance, optimization is typically not computationally feasible when there are dozens of continuous state variables – in such cases, numerical solution times are measured on the scale of years or centuries.[24]

A realistic model of a complicated economic phenomenon like the 2008 financial crisis requires understanding the behavior of many agents and many institutions, which require 'dozens of continuous state variables' to represent them. The reason that standard models fail is because they require calculating the *optimal* decisions for each agent. It can be difficult for real people to do this, and it becomes impossible even for a computer to do it when things get complicated.[25] The tractability problems of the standard modeling approach are unavoidable: It will never be possible to make successful models of complicated real-world situations that take the many interacting moving pieces of the economy into account.

In contrast, agent-based models don't try to find optima, so they don't suffer from this problem. In our housing model, for example, we considered as many as 100,000 households, all of whom differed from one another. We had not dozens, but hundreds of thousands of continuous-state variables. It would have been impossible to maximize each household's utility function while taking all the others into account. Instead, we mimicked the way that real households behave. At each step, each agent processed the information available to her and used simple heuristics to make decisions. Each agent's decisions consumed very little computer time, and as a result we could model 100,000 heterogeneous agents on a laptop. This will never be possible for a standard model based on utility maximization.

Agent-based simulation can easily incorporate realistic assumptions about the complicated structure of the economy. We can come closer to modeling the complexities of the real world, gaining verisimilitude in two different ways: The decisions of the agents can be more realistic, and the structure of the economy can be more realistic. But the most important consequence is that we can make *tractable* models, ones that produce solutions even when things get complicated. Whether or not direct models of behavior are more realistic than utility maximization (or its variants) will likely remain controversial for some time. However, the tractability problem of utility maximization for confronting complicated economic problems is not controversial.

This point was hammered home to me when I gave a presentation for the Macrofinancial Seminar at the London School of Economics in 2018. I was invited by Wouter den Haan, who is one of the leading European macroeconomists. After the seminar I met with a few expert micro-founded macro–modelers. To my surprise, they peppered me with procedural questions: 'How do you do this? How do you do that?' They were particularly keen to know how much effort it took to build and extend more complicated agent-based models. When I asked them why they were asking me all these questions, they began describing how much painstaking effort was required to add new features to their models.

In Chapter 5 you may have wondered why there are no micro-founded models that can simultaneously deal with the diversity of households and firms in a realistic way. The reason is that building complicated micro-founded models using rational expectations is hard, and one rapidly reaches a point where adding more realistic features becomes impossible. Agent-based models, in contrast, are only limited by the size and speed of the computers that run them. With a big enough budget for cloud computing, it is now feasible to build a model that includes every person and every firm in the world. Perhaps more importantly, it is possible to include all the relevant economic effects that cause the economy to change.

7.

Why Is the Economy Always Changing?

It's the economy, stupid.

James Carville, advisor to Bill Clinton (1999)

We can now return to the question raised in Chapter 1: What causes the economy to change? Essential measures of the economy, such as GDP, unemployment and inflation, fluctuate from quarter to quarter and year to year and show long-term trends driven by structural changes. These changes have a major impact on our well-being and, as James Carville famously noted, cause governments to rise and fall. Explaining such changes is one of the most important problems in economics. It is an important problem for everyone: The performance of the economy determines whether presidents are elected, and millions of jobs are gained or lost during the course of the business cycle.

Complexity economics and standard economic theory explain change very differently. As discussed in Chapter 1, there are two fundamental causes of economic change: outside influences and factors within the economy itself. With a few exceptions, standard economic models assume that the drivers of change come from outside.[1] I will argue that economic change results from a combination of internally and externally driven causes, depending on the context, and discuss how complexity economics provides natural explanations for this.

Spontaneous emergence of inequality and business cycles

Let's begin with a simple example of emergent behavior when we turn a standard model into a complexity-economics model.[2] In Chapter 5 we discussed the Solow model and how it was converted into the RCK model, which is a simple micro-founded DSGE model. The RCK model assumes a single *rational* representative agent who can calculate the optimal savings rate. What happens if we instead enter the wilderness of bounded rationality through the back gate, assuming boundedly rational (very stupid) heterogeneous agents? Does this fundamentally change the behavior described above, and if so, how?

We investigated these questions in a project led by Yuki Asano, whom I met at Oxford in 2017 while supervising his superb master's thesis in mathematics. He subsequently teamed up with Jobst Heitzig, Jakob Kolb and me to investigate modifications of the RCK model. The goal of Yuki's model was to illustrate a conceptual point rather than to build a realistic model. We turned the RCK model into a complexity-economics model by replacing the single, utility-maximizing, rational household of the RCK model with thousands of boundedly rational, heterogeneous households. Rather than calculating the optimal savings rate, the households simply copied one another, setting their savings rate using the two simple heuristics mentioned earlier, 'trial and error' and 'imitate the best'. The status-seeking tendency of households to copy one another has been well documented, so these seemed like sensible choices.[3]

We assumed that the households in Yuki's model live in a social network.[4] The nodes correspond to households, and a link indicates that they are neighbors, meaning they can observe one another's behavior. As in the original RCK model, each household saves a fraction of its income and consumes the rest. All the households work for the same firm and receive equal wages. In addition, each household contributes capital to run the business; the households pool their capital and invest it in the firm, and each household receives income based on the firm's profits in addition to its wages, proportional to its investment. As in real life, if a household gets richer it owns more of the means of production than poor households, and its higher investment profits allow it to consume more.

To get started, we randomly assigned each household its own savings rate, consisting of a number between zero (saving nothing) and one (saving everything). As in the RCK model, the households were allowed to adjust their savings rates. To accomplish this, from time to time a random household was allowed to update its savings rate by observing the consumption rates of all its neighbors. It then copied the savings rate of the neighbor who was currently consuming the most, with a small random error (hence using the heuristic of 'trial and error').[5] In more colloquial terms, each household looks at its neighbors, sees who has the fanciest car and biggest house, knocks on their door and asks them their savings rate, then begins saving at the same rate. Thus the agents in Yuki's model behaved like short-sighted, profligate consumers who want to 'keep up with the Joneses'.

A key parameter determining the behavior of the economy in the model is the typical time between updates of the savings rates of each household, which we called the *social-interaction time*. If the social-interaction time is small the household updates their savings rate frequently, and if it is large they do it only infrequently. When the social-interaction time is short, the economy gets stuck in a poverty trap, where the average savings rate and household consumption are very low. As the social-interaction time increases, the savings rate goes up and everyone gets richer. Remarkably, when the social-interaction time is long enough (meaning that savings rates aren't frequently updated), the savings rate approaches the golden rule, and the households become almost as rich as they would be if they were all perfectly rational. I say 'remarkably' because the agents in Yuki's model have *extremely* limited cognitive abilities; all they know how to do is compare consumption rates and select the savings rate of the highest consumer. If the social-interaction time is made even longer, the average savings rate remains near the optimal value. The fact that two simple rules of thumb can result in nearly optimal behavior is a good illustration of the effectiveness of heuristics.

Even more surprising, just as the social-interaction time gets long enough to make the savings rate nearly optimal, the economy suddenly starts to oscillate. The average savings rate and quantities depending on it, such as the output of the economy, make irregular fluctuations that resemble a business cycle, as shown in Figure 9. At the same time, the population spontaneously breaks into rich and poor groups. Households can move between the two groups: When there's a boom, there are more rich households; when there's a recession, the number of poor households increases. Interestingly, there is almost no middle class – nearly all households are either rich or poor.

How could such complex collective behavior arise from such simple heuristics? This was not obvious to us at first, and understanding it took considerable effort; emergent phenomena tend to be surprising by their very nature.[6] So if you find this hard to understand, you are not alone – it was hard for us too.

The key to the emergent oscillation is that the state in which everyone saves at the golden rule is unstable. Suppose that all households initially use the golden rule and have the same capital, so they have the same consumption. Now consider a household that makes

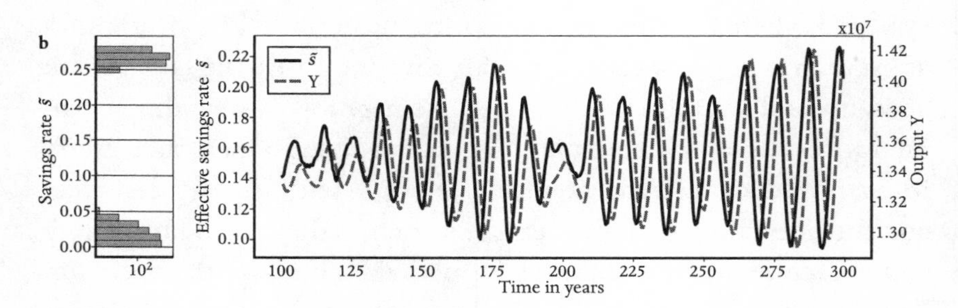

Figure 9. *An endogenous business cycle from a simple agent-based model.* When the social-interaction time is large enough, the average savings rate and the aggregate output of the economy spontaneously and irregularly oscillate. The panel on the left shows the relative number of households with each savings rate; poor households have savings rates below 0.05, and rich households have savings rates above 0.25, with almost none in between.

a small copying error that lowers its savings rate. This household will now consume more than the others, causing other households to copy it, and the evolving pressure will drive savings rates further down. But as capital drops and the number of poor households grows, the capital of the remaining thrifty households becomes more valuable, due to supply and demand: Scarce capital is worth more than abundant capital. So, despite their high savings rate, the thrifty households make such high profits that they also have the highest consumption. Other households then start to copy them, the average savings rate goes up and the cycle repeats itself.

Again, we didn't intend this to be a realistic model; we wanted only to demonstrate how the behavior of the economy can differ when a rational representative agent is replaced by many boundedly rational, heterogeneous agents. 'Imitate the best' and 'trial and error' are only two examples of many heuristics that describe the behavior of real households. Nonetheless, our work suggests that if we properly understood how households and firms really make decisions, we might find endogenous explanations for business cycles, and better understand and predict them. In fact, some of my complexity-economics colleagues have already made steps in this direction. There are now a few agent-based models that generate endogenous business cycles that look very realistic, and there are new agent-based models under development that aim to match business cycles for a given country at a given point in time.[7]

DSGE modelers are aware that rationality is an approximation. Real households are not rational, but their hope is that rationality (with appropriate frictions) is 'good enough'. My hypothesis is that the deviations from rationality are important and shouldn't be neglected. The planning errors that households and firms make are likely an important cause of business cycles – a cause that DSGE models, by their very nature, fail to account for. I believe that if we make more realistic models, in which households and firms are only boundedly rational, their planning errors will cause business cycles to emerge spontaneously (as they do in many agent-based models).

The question of whether bounded rationality plays a role in creating endogenous business cycles harks back to the debate from Chapter 1 about whether chaos is important in macroeconomics. As I mentioned there, during the 1980s macroeconomic models with chaos were proposed and then rejected, in part because it was necessary to assume unrealistic parameters to induce chaos. The unrealistic parameter in question was the aforementioned discount rate, which specifies the tradeoff between consumption now and consumption in the future. The representative agent in these old models was rational, and chaos happened only if the agent used a very large discount rate. The net effect was to make such agents myopic, in the sense that they cared much more about current consumption than future consumption. A theorem proven by the Brazilian-American economist José Shenckman in 1976 shows that a macro-model based on rational expectations won't become chaotic as long as the discount rate is reasonable.[8] This is called the *turnpike theorem*, evoking the image of a highway clearly visible to the horizon, where the future position of the car is easy to anticipate. The name suggests that an economy controlled by a rational agent who puts a sufficiently large weight on the future will ensure that the future is consistent with the present, avoiding chaos. In other words, in the absence of shocks, we should expect the model to settle into a static equilibrium. This theorem no longer applies, however, if the agents are boundedly rational. In this case, by definition, the agents cannot adequately anticipate the economy's future, and their planning may over- or undershoot the mark. The economy behaves more like a drunk driver on a winding mountain road, wobbling back and forth across the centerline.

Chaotic games

As mentioned in Chapter 1, my intuition about the prevalence of chaotic dynamics comes from Ruelle and Takens's remarkable proof. They showed that when a system is pushed sufficiently far from equilibrium, chaos becomes likely. Economic equilibrium is different from physical equilibrium, but – unless there's something special about economic systems – that proof is still relevant.

Ruelle and Takens's proof prefigured the inevitability of chaotic dynamics in turbulence; the work I am about to describe here attempts to do the same for the economy. (In Chapter 11, we'll discuss the analogy between fluid turbulence and financial turbulence, which makes this connection even stronger.) The turnpike theorem indicates that long-range rational planning in the economy suppresses chaos. Does this still happen under bounded rationality? If chaos does emerge under bounded rationality, under what conditions should we expect it to occur?

I began working on these questions in earnest some twenty years ago, when two Japanese postdocs, Yuzuru Sato and Eizo Akiyama, knocked on the door of my office at SFI and showed me some preliminary results investigating the game of rock-paper-scissors. I imagine most of you have played it: Two players simultaneously make a gesture, which can be either a clenched fist (rock), a flat palm (paper) or a V-sign (scissors). Paper beats rock, scissors beats paper, and rock beats scissors.

Studying rock-paper-scissors might seem silly, but games are widely used in economics to study situations in which agents receive payoffs that depend on their actions. Games capture the essence of strategic contests, and strategic contests play an important role in economics. In the stock market, for example, the movement of a stock's price depends on others' buying and selling, so your profits or losses depend not just on your decisions but also on those of everyone else in the market. Similarly, in a game, agents receive benefits that depend on their actions as well as the actions of other players. There are many different kinds of games: Rock-paper-scissors is what is called a *normal-form game*, in which the players make their moves simultaneously and get an immediate payoff that depends on a combination of their own move and those of the other players. Games provide an excellent testing ground for understanding how agents formulate strategies and how strategic competition evolves over time.

There have been five Nobel Prizes awarded in economics for game theory; the most well-known of the recipients is John Nash, protagonist of the 2001 film *A Beautiful Mind*, starring Russell Crowe. Nash is famous for his 1951 proof that normal-form games always have what is now called a *Nash equilibrium*.[9] This means that each player follows a strategy such that none of them have a better strategy unless one of the other players changes her strategy. In rock-paper-scissors, for example, the Nash equilibrium corresponds to both players randomly choosing rock, paper or scissors with equal probability. You can do this by rolling a die and choosing rock if the result is 1 or 2, paper if it is 3 or 4, and scissors if it is 5 or 6. If you play this strategy, no one can have a better chance of winning than you do. At a Nash equilibrium, all the players have settled into a strategy and have no incentive to change it. The Nash equilibrium may not provide the best outcomes for both players but, once there, the players are stuck. Just like economies, games can have multiple equilibria, in which case, even for rational players, it is not obvious which equilibrium will be chosen, and there may be suboptimal outcomes.

Nash equilibrium in a game is not the same as equilibrium in an economy, but it's a close cousin. At a Nash equilibrium all the agents have (locally) optimized their strategies, in the sense that there are no nearby strategies that are better. In economics, where the problem is to solve for prices and quantities traded, strictly speaking *equilibrium* means the point where supply equals demand. But game theory has become so integral to the way economists think about the world that they often loosely refer to *equilibrium* in the *strategic* sense as the set of agent decisions that (locally) optimize all their utilities.

When playing a game, a rational player will find a strategy corresponding to a Nash equilibrium, and in an economy, rational agents will settle into a set of decisions that maximize their utility. But when the players are *boundedly* rational, they may never settle on an equilibrium at all. If so, what happens? Can the behavior of the players become chaotic?

Eizo and Yuzuru were interested in these questions. To investigate them, they simulated the game of rock-paper-scissors on the computer with two players who each used reinforcement learning to formulate their strategies.[10] Under reinforcement learning, each player updates the probability of each possible move to favor moves that have been successful in the past. Eizo and Yuzuru hypothesized that under certain conditions the resulting dynamics are chaotic, never settling into an equilibrium. I collaborated with them in proving that this was true.[11] We showed that in the game of rock-paper-scissors, under reinforcement learning, most of the time the players don't settle into a Nash equilibrium; instead, they chase each other around the strategy space, as each tries to anticipate what the other will do. If one player starts to play rock more often the other will start to play paper more often, which will induce the other player to play scissors more often, and so on. The surprising thing is that most of the time this doesn't happen in a regular and predictable way – the dynamics are chaotic, meaning that the strategies never settle down and remain constant, and it is inherently impossible to predict what the players will do in the far future.

As a cultural aside, our work brought us into contact with the World Rock-Paper-Scissors Association, which promotes the game of rock-paper-scissors and organizes tournaments ('Dreams Do Come True! Play Rock-Paper-Scissors Professionally.').[12] They were enthusiastic when they heard about our paper, which coincided with what they observed. No one actually plays the Nash strategy of randomly making each move with the same probability. No one actually uses reinforcement learning either, but they thought the chaotic patterns we found, in which players chase each other's strategies, were closer to what they observed.

Rock-paper-scissors is just one of an infinite number of possible games. We provided a proof of principle for chaos in game theory, but maybe rock-paper-scissors is an anomaly? Rock-paper-scissors is special because its Nash equilibrium is purely defensive, meaning that, on average, it can never beat another strategy; in most games, the Nash equilibrium, on average, beats other 'nearby' strategies.[13]

When would chaos be likely and when would it be rare? My hypothesis was that, because complicated games – games involving many players and/or many possible moves – are difficult to learn, boundedly rational players would be unlikely to converge to a Nash equilibrium. Conversely, in simple games, converging to a Nash equilibrium should be likely. Tic-tac-toe is a good example of the latter. When I was nine years old, my friends and I had a lot of fun playing tic-tac-toe, but then I learned the strategy whereby the second player can always get a tie, and my friends quickly learned it too. The strategy we learned was the Nash equilibrium, and it made the game boring, so we stopped playing. Go and chess, by contrast, have many possible moves, and they're so complicated that no one has been able to discover their Nash equilibria. Nash proved there is at least one, and there may be millions of them – but so far no one has found any of them.

I had some ideas about how to prove or disprove my hypothesis, but after Eizo and Yuzuru returned to Japan, I lacked good collaborators. Then, about a decade later, while I was briefly living in Berlin before taking up my appointment at Oxford, I attended a session at the German Physical Society of presentations by physicists who were working in social science. I heard a talk by a bright young postdoctoral fellow named Tobias Galla, who was studying the multiplicity of Nash equilibria in games that were constructed at random. I approached him after his talk and we began a collaboration.[14]

Our idea was simple: We made up games at random, compiling a list of all possible combinations of players' possible moves and assigning a random payoff to each player for each combination of moves. To understand how we did this, think about a game with two players, Alice and Bob, and two possible moves, Up and Down. There are four possibilities: Alice and Bob both play Up, Alice and Bob both play Down, Alice plays Up and Bob plays Down, and Alice plays Down and Bob plays Up. For each of these four possibilities, we would randomly assign payoffs to Alice and Bob. However, we did this for games with more than two possible moves (where there is an explosion of different possible combinations as the number of possible moves increases). This procedure allowed us to generate any possible normal-form game, i.e. any game with simultaneous moves. Note that once we assigned the payoffs, they were fixed, so randomness entered only in creating the game – after that the game never changed.

For each randomly constructed game, we repeated the process that Eizo and Yuzuru had developed for rock-paper-scissors: We simulated the learning dynamics of the games using reinforcement learning and recorded what happened. Did the system settle into a fixed-point equilibrium, or were the dynamics chaotic? While it's impossible to investigate every possible game this way, by sampling the space of possibilities we got a good idea of the lay of the land.

The games we constructed weren't actually completely random: We could control the extent to which the game was competitive, as opposed to non-competitive. In a competitive game, if one player wins, the other player is likely to lose; the extreme case is called a *zero-sum* game, meaning that the winning player's gain is equal to the losing player's loss. In a non-competitive game, by contrast, if one player wins, the other player is more likely to win. We studied games that ranged from perfectly competitive (zero-sum) to perfectly non-competitive (meaning the payoffs for the two players are identical).

The set of all possible games is enormously large, so to narrow down the space of possibilities we decided to restrict our attention to two-player games with a large number of moves.[15] We showed that the likelihood of chaos depended on the degree of competitiveness and the way in which reinforcement learning is done. Reinforcement learning has a key parameter that determines the influence of recent moves relative to moves in the distant past; like discounting, it's usually better to pay more attention to recent moves. The details of our results don't matter here, but our overall conclusion for competitive games was that, when the reinforcement-learning parameter was set to sensible values, the learning dynamics were overwhelmingly likely to be chaotic. That is, the players never coordinated on an equilibrium; instead, they chased each other around the space of possible strategies in an unpredictable manner. Furthermore, the chaos was often *complicated* chaos, meaning that the process was effectively random and therefore difficult, if not impossible, to learn well enough to improve our reinforcement-players' predictions. We also observed other interesting things. For example, there were fads. For a while, the two players would fixate on a few possible moves, ignoring all the others. Then they would suddenly abandon those moves, replace them with others, play those for a long period, then shift again. This behavior was intermittent – there were times when there were steady payoffs and other times when payoffs fluctuated a great deal. In financial markets, this is called *clustered volatility*; in Chapter 11, I will argue that this correspondence is not an accident.

A key point our model illustrated is that equilibrium is an emergent property. Whether or not boundedly rational players will arrive at an equilibrium depends on how competitive and complicated the game is. This is not true for rational players, who will always arrive at an equilibrium where their expectations are consistent with the outcomes. When the games were sufficiently complicated and competitive, our boundedly rational players didn't do this – it is precisely the errors in the agents' boundedly rational expectations that cause the chaos.

A few years later, my graduate student Marco Pangallo studied the behavior of simple games with two players and only two moves. Was chaos possible in that case? He did a nearly exhaustive study, including an analysis of many variations of reinforcement learning and its relatives, and found that chaos in these circumstances is rare.[16]

We now understood games with an enormous number of moves, where chaos is common, and games with only two moves, where it isn't. What happens between these two extremes? Marco set to work. He broadened the scope of the investigation beyond reinforcement learning to include five other learning algorithms and found that, while there were some differences, the frequency of chaos in these algorithms and the circumstances in which it occurred were similar, implying that the behavior was robust. When he chose sensible versions of bounded rationality that capture key aspects of people's decision-making, the results were more or less the same. Then Marco invented a brilliant method to analyze the behavior of games with any number of players, using methods from statistical physics. He was joined by a clever German postdoc of mine, Torsten Heinrich. Together they applied these methods to understand the frequency of chaos for all possible normal-form games, depending on the number of possible moves. This showed that for competitive games the fraction of chaos increases rapidly as the game gets more complicated. The conclusion was clear: Games converge to a static equilibrium if they are non-competitive *or* simple; conversely, if they are both competitive *and* complicated, they are unlikely to converge – instead, their dynamics will be chaotic.[17]

These results have profound conclusions for economics. Agents often compete with each other in complicated economic settings, like a stock market or an industry with many competing businesses. The standard way economists model such settings is to assume equilibrium from the outset. Under rational expectations, the result is typically a fixed point, resulting in an economy that, as noted, changes only when it's disturbed from the outside. Our results suggest that, given the more realistic assumption that the agents are boundedly rational, one should instead expect chaotic dynamics. In a macroeconomic setting this means endogenous, irregular business cycles, like the ones observed in Yuki's model. Our results imply that standard economic models in any complicated, competitive setting are very likely wrong. This shows that we need complexity economics to understand complicated, competitive phenomena like business cycles.

PART III:
The Financial System

8.
Inefficient Markets

Observing that the market was frequently efficient [efficient market theory adherents] went on to conclude incorrectly that it is always efficient. The difference between these propositions is night and day.

Warren Buffett[1]

What does the financial system do?

The essential function of the economy is to structure and coordinate our work, and allocate resources, including human capital. In a thriving economy, individuals are far more effective than they would be if we had to survive on our own. The financial system plays a subtle but essential role in directing this coordination.

If the economy is the metabolism of society – its 'digestive tract' – the financial system is analogous to the brain in our gut, which helps us digest food. This is formally called the *enteric nervous system*, but is also nicknamed the *gut brain* or the *second brain*. It stretches from the esophagus to the rectum, and has half a billion neurons, five times as many as in the spine and almost as many as the entire nervous system of a cat. It directs the activities of the digestive system while operating almost entirely independently of the brain in your head. If your gut brain ceases to function, you lose the ability to digest food, and, without intravenous feeding, you die. The digestive system is just another name for our metabolism, which the gut brain guides and controls. Remarkably, even though we can't live without it, very few people have ever heard of the gut brain.

Analogously, the key function of the financial system is to guide and control the economy by determining what we will or will not produce. It plays an important role in processing information and making allocation decisions. Like the gut brain, the financial system (in a capitalist economy) operates more or less autonomously, though it needs regulatory guidance in order to work well. How much regulation it should have, and what kind, is hotly debated.

Shipwrecked on his desert island, Robinson Crusoe could consume only the fruits of his own labor – for him, money would have been useless. But as Adam Smith pointed out, in a modern economy we are able to specialize, and we do this by using money to consume the fruits of the labor of others and to allow them to consume ours. The financial system does its job by controlling where money goes. Fundamentally, money *directs* what we work on and what we produce. Someone says, 'Hey, I've got a good idea for a product and the competence to bring it to market,' and a rich person responds, 'Okay, I believe you, and if you'll give me a cut of the profits, I'll give you some money.' Then a team of people goes to work with a common purpose. In an ideal world, people who work in the financial system provide a valuable service by processing information and making decisions so that effort and resources are allocated where they can be most effective.

Contracts are an essential tool of the financial system. They operate by structuring the flow of money and activity over time. Ownership of a stock or a bond is a contract, relating an investment at one point in time to payments at other points in time. Other contracts include loans, mortgages, employment agreements, insurance policies, patents, pensions and wills. Contracts specify the commitments of parties to one another and synchronize their activities. They can also transfer risk – for example, through insurance policies. If an airline wants to lock in the price of jet fuel, it can buy a contract allowing it to transfer the uncertainty about future prices to someone else. Contracts coordinate our actions by making future commitments and dependencies clear.

Contracts form a vast web.[2] They are an important component of the network of balance sheets, connecting us in space and time. Each contract sits on someone's balance sheet, either as an asset or a liability. The relationship is reciprocal; my liabilities are someone else's assets. The insurance policy on my house says that if it burns down, money flows from others (who may be far away) to me, so that local people can rebuild my house.

As discussed in Chapter 2, economics can be viewed as a combination of accounting and human decision-making. Contracts are an essential part of accounting, which is a mechanical process following a set of simple laws.[3] The web of contracts is like a set of wound springs linking us all together; as the conditions in the contracts are triggered, the springs unwind and prescribed actions take place, affecting our decisions. If we could know the web of all contracts in detail, we could make powerful predictions about the global economic system from this information alone.

The financial system provides a mechanism for processing information and making decisions in the economy. The way it does this is subtle, generating debates about how well it functions. The *efficient-markets hypothesis* assumes that the system functions perfectly. My theory of *market ecology*, described in Chapter 10, offers a complementary view. According to that theory, the *inefficiencies* of the financial system are fundamental to its operation; admitting that markets are inherently inefficient, and understanding what causes these inefficiencies, is essential to understanding why markets can malfunction.

The relative size of the financial system in the economy has grown enormously in recent history. Shortly after World War II, the US financial sector accounted for 2.8 per cent of GDP; by 2006, it was 8.3 per cent.[4] In 1980, a typical financial-services employee earned about the same wages as her counterpart in other industries; by 2006, employees in financial services earned an average of 70 per cent more.[5] Between 1969 and 1973, only 6 per cent of Harvard undergraduates went into financial services after graduation; by 2008, this had grown to 28 per cent.[6] This suggests that the financial system is directing an increasing amount of money and effort toward itself rather than toward the rest of the economy.

Is the increase in resources devoted to the financial system necessary because the economy has become more complex, requiring more information to make sound investments? Is this increased effort stimulating more innovation or increasing the efficiency of production? Or are we suffering from an ailment where our collective gut brain has gone rogue, and is engaging in cancer-like activities rather than doing its job?[7] Perhaps major components of the financial system do nothing more than act as a global casino for the very rich, while destabilizing the economy with little positive benefit to the economy as a whole? In this part of the book, I will argue that a complex-systems point of view can help answer these questions.

The financial system is a vast complex system unto itself, with an array of interacting institutions. These include regulated entities like commercial banks, investment banks, insurance companies, mutual funds, pension funds, hedge funds, private equity funds, venture-capital funds, housing markets, stock exchanges, central clearing parties, security and investment dealers, and special-investment partnerships – as well as a host of other entities that constitute the shadow banking system, which creates credit without regulatory oversight. However, the Fed's FRB/US model and its counterparts throughout the world use simplified models of the financial system that lump most (usually all) of these diverse institutions together. The complexity-economics approach, in contrast, can take the differing role of each type of institution into account, in order to better understand how the financial system functions, and explain why financial crises occur and what we can do to fix them.

Beating the stock market

In 1985, while I was an Oppenheimer Fellow at Los Alamos, I began mentoring my second graduate student, John Sidorowich (universally known as Sid). Sid and I worked on a research project that merged my earlier collaboration on predicting roulette with the Dynamical System Collective's work on identifying chaos, as described in Chapter 1. We took advantage of the fact that chaos is a double-edged sword: On one hand, chaos imposes fundamental limits to long-term prediction; on the other, it means that data that otherwise look random can be predictable in the short term.

There are many situations where useful models can be built even with no fundamental understanding of causality. The best predictions of sunspots, for example, are made by fitting a simple mathematical model to historical records of sunspot levels.[8] The work I did with Sid showed that if random-looking behavior is caused by simple chaos, one can take advantage of this to make better predictions than those made using standard time-series (historical) models. Our research led to a 1987 paper called 'Predicting Chaotic Time Series'.[9]

Our method made good predictions as long as we had enough data and the mechanism underlying its randomness was simple chaotic dynamics. Some of the phenomena we predicted included weakly turbulent fluid flow, sunspots and ice ages. I frequently spoke about this work at seminars and conferences and, when it came time for questions, someone would inevitably raise their hand and ask, 'Have you tried applying your methods to the stock market?' I knew nothing about finance, but the question came up so often that I couldn't help thinking about it. I got tired of answering, 'No.'

By 1991, I had been at Los Alamos for nearly a decade. I had great collaborators and really liked my complex-systems group. But I was spending too much time on administration, fund-raising and dealing with government bureaucracy. What to do instead? I kept reflecting on the question that everyone always asked about my work on prediction. *Was* it possible to beat the market? At that time, the debate was dominated by Eugene Fama, a professor in the business school at the University of Chicago who would go on to win the 2013 Nobel Prize in economics. Often regarded as the 'father of modern finance', he argued strongly in favor of the efficient-markets hypothesis, which says that getting an advantage by predicting stock prices is impossible.[10]

The efficient-markets hypothesis flows automatically from the assumption of rational expectations. If investors are rational, then stock market prices should fully and accurately reflect all available information, and prices should change only when new information enters the market. New information is by definition unpredictable (or it wouldn't be new); ergo, future changes in prices should be random. Fama and most other financial economists at the time believed that trying to predict stock prices based on publicly available information would be pointless. I didn't think this could be true, and I liked the idea of disproving it.

Fama was not the first to suggest that stock-price changes are random. In fact, financial markets were the original inspiration of the random-walk model, introduced by the French mathematician Louis Bachelier in 1900. In his PhD dissertation, he proposed that the price of a stock today is the price yesterday plus a random number, and the price tomorrow is the price today plus a different random number, and so on.[11] This is like a drunk who randomly steps to the right or to the left, with no bias toward moving in either direction. If the drunk starts at a tavern door, where is he after 100 steps? Knowing the size of his steps, and assuming that steps to the right or left are equally likely, we can compute the probability that the drunk will be at any given distance from the tavern. But he is equally likely to end up on the right or the left; we cannot predict the direction of his movement. Similarly, if stock prices follow a random walk, we can estimate how far prices will drift from current prices after a given time, but not whether the change in price will be up or down.

Beating the market seemed like a good challenge, so in 1991 my childhood friend Norman Packard and I teamed up again and recruited another old friend from Santa Cruz, Jim McGill, to run the business side of things. We started Prediction Company, which in hindsight was one of the earliest successful quantitative hedge funds. We made a deal with the options-trading firm O'Connor and

Associates, which was eventually acquired by Swiss Bank Corporation, which itself later merged with the Union Bank of Switzerland (UBS). They provided us with all the investment capital we needed, and we got a cut of the profits; our job was to find and implement a good trading strategy. For us, the stock market was like a casino, only the stakes were much higher, you didn't have to hide your computer, and they didn't throw you out or break your kneecaps for winning.

At the time we knew very little about finance or economics, so we treated the stock market as a vast stream of numbers. Our strategy was to find 'pockets of predictability' in historical prices that recur more often than they should at random; we would then make bets based on these patterns. This book is not about beating the market, so I won't go into detail about what we did. Nonetheless, our systematic approach, which allowed us to beat the market, provides some valuable methodological lessons that are relevant to making better models of the economy.

Data and infrastructure were the foundation of everything we did at Prediction Company. We collected all existing historical data on US stocks. This included not just prices and trading volume but options, insider-trading reports, analyst estimates and anything else we could lay our hands on. We hired a few really good computer-programmers and developed a framework that let us automatically process vast amounts of data using an array of Sun workstations, which were the most affordable powerful computers then available. We created a special-purpose language for processing financial data and developed a database program to quickly process prices and other data. The resulting infrastructure made it easy to build models and simulate trading.

Using machine learning and other statistical methods, including a few we invented ourselves, we discovered several predictable patterns, but we also realized that there was a great deal of useful information in the published literature in finance – particularly in papers claiming to find violations of efficient-market theory. For example, such a paper might say, 'We tested trading rule X on stocks in the S&P index between 1970 and 1991 and showed that it produced returns 5 per cent in excess of those of the S&P index.' Because of the strong bias toward efficient markets, papers like this were hard to get published, but some still managed to make it into print, mostly in second- and third-tier journals. Our powerful framework allowed us to rapidly and easily test whether the results of papers like this were correct. We systematically tested the results of every published paper that claimed to find deviations from randomness in stock markets.

The results were sobering, suggesting that, at least on this topic, the published work was not reliable. For around half of the papers, we couldn't reproduce the results, even when we tested the postulated deviation from efficiency using the same data. The results in another fourth of the papers were correct when we used the same data but failed when we tested the hypothesized method using different time periods or stocks. Fortunately, the remaining fourth we tested contained useful, reproducible results.

Nonetheless, when taken by themselves, none of the trading rules in the validated papers produced the steady profits we were seeking. Even if the overall returns were good on average, they were uneven; the risk was too large for our weak stomachs. But when we combined the trading rules from all of the validated papers with those we found on our own, the result was an effective trading system. As a result, Prediction Company made substantial returns that were about five times steadier than those of a typical mutual fund or stock index.[12]

When we started Prediction Company, Eugene Fama was quoted in a popular science magazine as saying that the odds of our success were 'not zero, but they are very, very low'.[13] This was a cute way of saying that, even though stock prices are random, there was always a small chance that we'd get lucky. The metaphor Fama and others liked to use was of monkeys with typewriters: If enough monkeys type long enough, one of them will eventually duplicate Shakespeare's complete works.

A statistical analysis of our trading results makes it abundantly clear that our steady profits were not due to chance. Prediction Company ran for twenty-seven years, with only one losing year – 2007.[14] The odds of achieving Prediction Company's track record at random are vanishingly small. Our performance showed that, with the right information and the right model, one can take advantage of hidden patterns in stock prices. We proved that, strictly speaking, the efficient-markets hypothesis is wrong. And we weren't the only ones; although we were among the first, a few other firms did even better.

After eight years at Prediction Company, I was eager to get back to basic research. In the process of becoming an expert on the stock market, I had also begun to read more widely about finance and economics and had concluded that economics needed new ideas. While mainstream economics had flirted with chaos and complex systems, neither had taken hold. This created an opportunity to do something new.

At Prediction Company, we built successful models that made good predictions, but the models were black boxes that didn't explain the mechanisms that caused the inefficiencies we exploited. We had to keep our models secret, which meant they contributed little or no value to society as a whole. I wanted to build models that could explain *why* markets behaved as they did – models that anyone could use and that would help solve the world's problems. This led me back to the Santa Fe Institute, where I began doing basic research on financial markets.

How well does the market do its job?

Let's revisit the question of how well markets do their job. As discussed, financial markets are supposed to help us allocate our efforts so that they're productive. But what does beating the market have to do with the way we allocate effort?

The phrase *market efficiency* has two senses. The first, 'allocative efficiency', means that prices are always set correctly, so effort is allocated appropriately. Stock prices help people decide how to do that: If the price of pork bellies goes up, more farmers will grow hogs. The second, 'informational efficiency', means that you can't increase profits by making better predictions about stock prices. If all investors are rational and operate within perfect markets, everyone's utility is maximized, and the market is efficient in both the allocational and informational sense.[15] In this idealization, the statement that the market casino cannot be beaten is equivalent to saying that resources are allocated correctly. However, this depends on strong assumptions that aren't well-supported in practice. Under more realistic assumptions – for example, when investors are boundedly rational – these two different notions of efficiency are no longer equivalent. In fact, as I will argue, they can be really different.

To better understand the logic behind the efficient-markets hypothesis, let's make an analogy to traffic. Suppose there are two possible routes from City A to City B. Without traffic, the travel time for each of the two routes is the same, but if there is traffic, one may be faster than the other. If everyone has access to traffic reports and chooses the fastest route, then the faster route will gain more traffic until the travel times adjust so that they're the same. If this happens everywhere, the traffic system is efficient in the informational sense, because no one can do better by picking a faster route, and in the allocative sense, because cars are using the road system as well as possible to keep everyone's travel times as short as they can be. But there's a built-in contradiction, because if everyone believes that everyone else is processing the information needed to choose the best route to make the system efficient, then all the routes are the same, and there is no incentive for anyone to process information, which will make the system inefficient.

A famous joke illustrates the paradoxical nature of the efficient-markets hypothesis. Eugene Fama and a friend are walking down the street when they see a $100 bill. Fama says to his friend, 'There's no point in picking it up. It must be counterfeit, or someone would have picked it up already.' To this I can add a true story: About two years after we formed Prediction Company, Norman and I left our office on Aztec Street in Santa Fe to go to lunch. We looked down and saw a $50 bill. We picked it up and used it to buy an especially nice lunch at our favorite New Mexican restaurant. The money may have been counterfeit, but the lunch was delicious.

The introduction of Google Maps shows how efficiency depends on technological progress. By providing useful information about travel times, Google Maps has substantially altered our driving habits and made the road system more efficient.[16] Similarly, in the next section we will see how computer technology moved the reference point for market efficiency in finance.

Econophysics, market impact and the usefulness of efficient markets

The efficient-markets hypothesis has been controversial since its inception, but while it is no longer the iron-clad pillar it once was, it is still one of the central ideas in mainstream financial economics.[17] Our experience at Prediction Company and that of other successful trading firms shows that, strictly speaking, the hypothesis is wrong. However, this doesn't mean it isn't a useful approximation for some purposes. There are situations where it works very well and others where it fails miserably. The efficient-markets hypothesis is a good approximation for understanding some aspects of markets when they are functioning properly. At the same time, it is manifestly wrong: It predicts that markets never fail, which defies both experience and common sense. We'll explore some examples that illustrate both points, beginning with some success stories.

The first is simple: The theory of rational expectations, which predicts that the market should be efficient, explains why it is hard to beat the market. While I am proud that we succeeded in beating the market, it wasn't easy; creating our first strategy making consistent, reliable profits took five years of hard work for me and twenty others at Prediction Company. This is a good example where market efficiency, while not strictly true (our results disproved it), is a reasonable approximation.

Another success story is *derivative pricing*. Derivatives are financial instruments whose value is based on other assets (their price is *derived* from those assets). The classic example is a stock option, which allows you to buy or sell a stock at a given price (called the *strike price*) at a given time. This can be very useful to a business: An airline that wants to ensure that rising oil prices won't cut too much into its profits, for example, can buy an option to purchase oil at a fixed price, no matter how high the market price rises.[18]

Mathematical methods for pricing options are based on the efficient-markets hypothesis. This is done by assuming that the underlying stock price follows a random walk. For example, suppose we want to estimate the value of an option to buy – a *call option*. We do so by computing how often and by how much the random walk for the underlying price rises above the strike price. The idea is simple: If the price of the underlying stock fails to rise above the strike price, the value of the option is zero; but, in cases when it surpasses the strike price, its value is the price that was reached minus the strike price. The value of the option is then computed by averaging all possible realizations of the random walk. (This idea was introduced at a practical level by Edward Thorp and Sheen Kassouf and refined by Fischer Black, Robert Merton and Myron Scholes.[19])

A much less well-known use of the efficient-markets hypothesis is for deriving the shape of the *market-impact function*, which describes how changes in supply or demand affect price. When a trader decides to buy an asset, the demand for that asset rises, which tends to drive its price up. The market-impact function quantifies the average increase in price in response to initiating a trade of a given size. Remarkably, the market impact function has a universal shape – that is, it's the same in all markets where it's been tested: stocks, bonds, commodities and currencies.

I have chosen this example for several reasons: market impact is practically important; its universal shape illustrates a quantitative 'law' in economics, much like those in physics; and the theory that was developed to understand it illustrates how it can be useful to enter the wilderness of bounded rationality from both gates.

First, its practical importance. At Prediction Company, we worked hard to understand our market impact because, since you make money in markets by buying low and selling high, it lowered our profits and was the gating factor on how much money we could successfully manage.[20] Market impact increases with trading size, meaning larger orders are bought at higher prices and sold at lower prices than small orders. This limited the amount we could profitably buy and sell; if we tried to trade too much money, the market impact became too large and our profits went to zero.

Prediction Company was among the first to carefully measure market impact. We did this by recording the price just before our orders hit the market and just after they were executed. We anticipated that market impact would go up linearly with order size, or perhaps even faster than linearly. In fact, we found just the opposite: It increased at a slower-than-linear rate. In other words, small orders had a larger market impact per share than large orders, so that the shape of the curve describing the shift in price became progressively flatter with increasing order size. We found that the shift in price increased roughly as the square root of the size of our trade (a fact that has now been extensively confirmed).[21]

This was great news for us, because it allowed us to trade at larger volume and make more profits than we would have expected. But it was also a puzzle – why should market impact look like this? Furthermore, though our overall impact in heavily traded assets was lower than in lightly traded assets, the curve had the same square-root shape for all assets. Why should this be?

At Prediction Company, we didn't have time to pursue questions purely for their own sake. We just accepted that market impact was roughly described by a square-root function and used this to optimize our trading strategy. But after returning to basic research, I was keen to understand why this was true.

To understand why the universal shape of market impact is so remarkable, let's revisit the law of supply and demand, which is perhaps the most important and most reliable principle in economics. The law of supply and demand says that prices tend to move to the place where supply equals demand, so if supply goes up, prices go down, and if demand goes up, prices go up. It's a powerful and useful law, but there is no good theory explaining the shape of supply and demand curves. When Professor Gurley sketched the curves on the blackboard in my Economics 1 class at Stanford, as mentioned in the introduction to Part I, the price could be any increasing function of demand and any decreasing function of supply. Restated in physics jargon, there is no universal functional form.[22] While economists derive the shape of supply and demand curves from economic theory, this depends on the utility function chosen for the agents – different utility functions lead to differently shaped curves. But there are many possible utility functions, with no good empirical justification for one over the other, so even if everything else in the theory is right, the result can at best be qualitatively correct.

Market impact is closely related to supply and demand – it indicates how much the price changes when demand or supply changes. But if the supply and demand curves can take any form, how could the market impact be the same in many different markets? This seemed to be an important clue about something fundamental to economics.

The story of how we came to understand market impact intertwines with my own personal narrative. It involves a movement called *econophysics* that emerged in the late 1990s, around the time I left Prediction Company. During the latter half of the twentieth century, finance had become increasingly mathematical. The job market in physics had grown tight in the late 1960s, and, because physicists are good at math, more and more of them were hired on Wall Street to work on tasks such as derivative pricing. Econophysics took this a step further, as physicists began doing basic research in economics – writing and publishing papers in academic journals rather than just doing applied research for hire.

The arrival of physicists in economics, as part of the econophysics movement, was met with some resistance by economists. The physicists weren't just trying to help economists with their math problems; they were suggesting conceptual frameworks that were often contrary to those of economists. The new assumptions and ways of thinking that physicists wanted to import into economics were frequently antithetical to economic orthodoxy. The economists felt under attack, and they weren't going to let the physicists into their club, so if physicists wanted to be part of a research community, they needed to create one for themselves. Two of the leaders in creating this movement were H. Eugene Stanley, a very distinguished physicist from Boston University, and the Sicilian physicist Rosario Mantegna. In 1999 they wrote a book called *Econophysics*, and the name stuck.[23]

There are several ways in which physicists approach science differently than economists do. Physicists are taught to stay as close to the data as possible; theories should emerge from empirical observations. Economists, on the other hand, are taught to build theories and then test whether they match the data. Their justification is that, unlike in physics, economic data is so noisy that one needs a theory to make any sense of it.[24]

Another important difference is that physicists believe that theory must ultimately seek to be quantitative, in a strong sense. It is not enough to say that matter attracts other matter, as Descartes did; the theory of gravity became useful only when Newton asserted that the gravitational force decays inversely with the square of distance. When combined with Newton's other famous law, $F = ma$, this allowed him to predict the motion of planets and comets far into the future.[25]

This is why physicists like me got so excited about the possibility of a universal law for market impact. It suggests that there is a quantitative law for changes in supply and demand in economics, a bit like the inverse-square law of gravity. This was doubly exciting because market impact is a kind of force. From a trader's point of view, market impact is like friction – it bleeds away profits. But, from the point of view of a complex-systems scientist seeking to understand how markets work, it is an interaction rule: Trading changes prices via market impact, and changes in prices cause trading, and so on. (This runs directly counter to the mainstream view, which holds that price movements are strictly due to new information.) Market impact relates to how quickly prices approach equilibrium, so it is an important clue about what happens out of equilibrium, partially answering one of my original questions to Professor Gurley.[26]

When I left Prediction Company to rejoin SFI, one of my goals was to understand the square-root law of market impact. I thought it would be useful to look at the way trades are actually made. The most common way to organize trading in modern markets is via what is called the *limit-order book*. Market participants can place orders to buy or sell at a given price or better. If these orders don't result in immediate transactions, they are placed in the limit-order book, which has a list of all the active quotes.

Most theories of financial markets are based on rational investors. Inspired by some earlier work, I decided to do just the opposite and assume that people put orders in the limit-order book and canceled them at random, except for the fact that everyone keys off the current price.[27, 28] What would happen in that case? Would we recover the square-root law of market impact? I teamed up with several SFI collaborators, including Marcus Daniels, László Gillemot and Giulia Iori, and the effort was led by our brilliant colleague Eric Smith.[29] The zero-intelligence assumption allowed us to explain several interesting properties of limit-order books using powerful mathematical methods from statistical physics.[30] This led to some strong, empirically testable predictions. For example, an important property of the market is the *bid–ask spread*, which is the difference between the highest-priced buy order and the lowest-priced sell order in the limit-order book for a particular stock at any point in time. We compared our model to real data and were able to accurately predict the bid–ask spread based on rates of order placement and cancellation.[31]

There were, however, parts of our model that didn't fit. We were disappointed that we didn't reproduce the square-root law of impact very well. While the impact function had the right shape – the slope decreased with order size – it changed depending on assumptions about the flow of buy and sell orders into the market, and it often deviated substantially from a square-root function.

And there was something else that we couldn't explain: Even though the agents in our model placed random orders, the resulting price movements weren't random. Rather than making a random walk, the price movements were *mean-reverting*, meaning that, if the price went up at one point in time, it was more likely to go down at the next point in time. Rather than being equally likely to step left or right, the drunk tended to reverse himself – if he stepped right, his next step was more likely to be to the left. This was puzzling. We had expected random order-placement to yield an efficient market, but instead it yielded a market inefficiency, that is, a predictable pattern in the prices.

This problem was solved in an elegant way by fellow econophysicist Jean-Philippe Bouchaud and his collaborators. Jean-Philippe is both a strong mathematician and a brilliant statistical physicist, who also runs a successful hedge fund. He realized that our model showed that order-flow could not be random. If it were, it would create an *arbitrage*, meaning that a clever trader could make money from the non-random movement in prices. He and his collaborators demonstrated that the assumption of market efficiency implies the square-root law of market impact.[32] Or, said differently, imposing the compatibility of a purely random order-flow with price unpredictability mechanically leads to a square-root impact law.

The upshot is that market efficiency leads to a universal quantitative law for changes in supply and demand, much like the laws of physics. To summarize: In order to be consistent with the operation of the limit-order book, the shifts in prices caused by placing orders must increase as the square root of order size – otherwise there is an inefficiency that can be exploited by arbitrageurs.

In Chapter 6, we discussed the metaphor of exploring the wilderness of bounded rationality. Mainstream economists like to enter the wilderness through the front gate of rational expectations, while complexity economists like to enter it through the back gate. Both seek to understand the bounded rationality of real people that lies somewhere in between, deeper in the wilderness. The story of how we explained the square-root law of market impact is a good example of how it can be useful to first enter through the back door, then use the front door.

Random order-placement is an example of a *zero-intelligence* model – the exact opposite of rational expectations – an idea introduced by the economists Dhananjay Gode and Shyam Sunder.[33] While zero-intelligence models are a bit far-fetched – people are at least somewhat intelligent – they can provide a very useful starting point. If agents have no intelligence, then they are like atoms in physics. This is what allowed Eric Smith, who was well-trained in the branch of physics called *statistical mechanics*, to construct a nice theory for what goes on inside the limit-order book, which led to simple formulas allowing us to make empirically testable predictions.

The term *zero-intelligence* makes people laugh: When a summary of our paper was translated into Russian, it erroneously said that our main result showed that 'traders are complete idiots'. The Russians thought this was so funny that it was posted on the leading Russian website for jokes, where it got a huge number of hits. Nonetheless, by exploring the consequences of agents with no intelligence whatsoever, we could better understand the dynamical properties of limit-order books. We showed that this market institution is not inert but, rather, has its own characteristic dynamics – suggesting that a little intelligence needed to be added to make a model whose predictions were consistent with reality. In this case, the assumption of market efficiency, which is a consequence of rational expectations, provided the extra element needed to get the right answer.

The slow and uncertain path to market efficiency

The opposite of market efficiency is market inefficiency, which translates into a persistent profit-making opportunity. The argument for efficient markets depends on the idea that buying drives prices up and selling drives them down, causing market inefficiencies to quickly disappear. An investor who processes information and weighs the merits of one stock against another and trades accordingly is called an *arbitrageur*. If Apple stock is underpriced, arbitrageurs will buy it and drive the price up until it's properly priced. According to received knowledge, their action should quickly remove any future opportunities to make profits, making the market efficient.

But as already noted, this argument is self-contradictory: Market efficiency requires arbitrageurs who are attracted by profits. However, if markets are perfectly efficient, then there are no profits to be made by arbitrageurs, and therefore no reason for them to participate in the market. This means markets cannot be perfectly efficient – perfect market efficiency is inherently paradoxical.

The advocates of efficient markets are aware of this paradox.[34] They respond that, although arbitrageurs make profits, these are small – big enough to motivate the arbitrageurs but small enough that market efficiency is still a good approximation. The real debate is about whether or not deviations from market efficiency are important. Are they big enough to make a difference?

The proponents of market efficiency also argue that clever investors rapidly discover and exploit any market inefficiencies and return the market to an efficient equilibrium. But, assuming this happens, how long does it really take? Our research at Prediction Company produced one of the few published results on this.[35] Our most successful strategy was something called *statistical arbitrage*. We built the strategy using eight years of historical data from the late 1980s and early 1990s. Then, when we suddenly acquired data going back to 1975, we were able to test our strategy on data over a twenty-three-year span. The results, which provide a sobering contradiction of the mainstream view, were really surprising.

As already mentioned, we built our trading strategy by taking advantage of many different types of market inefficiencies. We could tell when the market was behaving inefficiently because prices or other quantities behaved in a particular way, flagging that an inefficiency was present. We called these informational flags *signals*. One of our best signals was associated with what is called *convergence trading*. We would look for similar stocks, one of which was overpriced and one underpriced, and bet their prices would converge. Another signal used analyst estimates as its inputs – if analysts upgraded their estimates for a stock, this signal predicted it was likely to increase in price for a while, and vice versa.

Though we combined all the signals in our trading strategy, each of them could also act as a stand-alone trading strategy. When we measured the profits from these two signals over a long span of time, we saw that, though they changed in different ways, they were both very persistent. While our convergence trading signal made excellent profits throughout the 1990s, if we had been able to execute it in 1975, our analysis showed that we would have made *phenomenal* profits. Although the performance of this signal steadily declined over the twenty-three-year period by roughly a factor of 3, it was still going strong in 1998. Does this mean the glass was half full or half empty? An advocate of efficient markets would emphasize that the profits decreased with time, indicating that the market was becoming more efficient. True, but it was certainly taking its time in doing so. What we observed was not at all the rapid return to market efficiency postulated by proponents of the efficient-markets hypothesis.

The behavior of the analyst-estimate-based signal was even more striking. No one could have traded this signal until 1984, when a regulatory change made the data it needed available in a timely manner. This signal's profits were initially very low, but they steadily grew over the next fifteen years. This is exactly the opposite of what the theory of market efficiency would predict! The fact that the market became increasingly *inefficient* over time is a strong contradiction of the efficient-markets hypothesis.[36] How could this be?

Why is the path to market efficiency so slow? There are several reasons.[37] The first has to do with the challenge, in statistical analysis, of distinguishing the signal (the information useful for prediction) from the noise (the random background fluctuations that make it harder to detect the signal). To create a quantitative trading strategy, investors must identify weak signals in a noisy background. Even for good trading strategies, like those we used at Prediction Company, the signal is about twenty times weaker than the noise.

To create a reliable trading strategy we had to accumulate enough data to be able to pick the signal out of the noise – more data improves the accuracy of statistical tests. For signals like these, it takes around five years of data to effectively distinguish the signal from the noise. When you factor in the time it takes to detect the signal, develop and test the strategy, and convince someone to provide capital, it can take decades for a trading strategy to be discovered and reach maturity.

An advocate of market efficiency might say that the mere fact that the signal-to-noise ratio for trading signals is so small supports market efficiency. In a sense, I agree: It is challenging to beat the market. But huge profits result from small inefficiencies, and as I will argue in Chapter 10, this can lead to substantial market failures.

The second reason that market inefficiencies can be so persistent is technological progress. At Prediction Company, we used computers to trade 1,000 stocks every day, without human intervention. In 1975, it would have been impossible to implement our trading strategy with the computers and data-management technology available at the time.

Finally, as the market becomes more efficient, it approaches perfect efficiency more and more slowly. If the market ever managed to reach perfect efficiency, all trading signals would disappear. The closer the market comes to perfect efficiency, the weaker the signals become, which means they take longer to detect, making the continued approach ever slower. In a 2009 paper called 'The Reality Game', Dmitriy Cherkashin, Seth Lloyd and I showed that this is true, even in an idealized setting.[38] (To be more precise, we showed that the inefficiency decreases in time as a power law, which approaches zero much slower than an exponential.)

This explains why the approach to efficiency is so slow, but not why inefficiency can sometimes increase over time. The simple explanation is that the landscape is always changing. The participants in the stock market change. The rules change. New financial instruments, such as mortgage-backed securities, are introduced. Innovation takes time. New strategies often render old strategies less effective, but they can also make them more effective. Because market participants are only boundedly rational, the capital invested in a given strategy may grow too large and create instabilities. New inefficiencies continually emerge from the shifting landscape; it can take decades to detect them, and as a result, the market never settles into the efficient nirvana that Eugene Fama and others imagine.

Even small inefficiencies can have a big effect on how well markets function – small informational inefficiencies can cause large allocational inefficiencies. As we'll discuss in Chapter 10, to fully understand this process, we must think about market dynamics in ecological and evolutionary terms. But before we do so, let's look at some other ways in which market inefficiencies manifest themselves.

9.
The Self-Referential Market

October: This is one of the peculiarly dangerous months to speculate in stocks. The others are July, January, September, April, November, May, March, June, December, August, and February.

Mark Twain[1]

The efficient-markets hypothesis also has implications about how prices respond to outside news. Markets have endogenous dynamics – prices change on their own, even without news affecting fundamental valuations. Put differently, markets make their own news.

The 1987 stock market crash

1987 was an important year for complexity economics, for two reasons. First, it was the year of the SFI meeting on 'The Economy as an Evolving Complex System', mentioned in the introduction to Part I – a meeting that eventually led to the creation of two influential agent-based models that help us understand why prices behave as they do, and why market inefficiencies can cause market malfunction.

The second stimulus to complexity economics occurred later that year, on October 19, when the US stock market lost more than 20 per cent of its value in a single day. A drop in prices of this magnitude is huge: The stock market reflects the total worth of the biggest companies, whose value is a substantial share of the US economy. If you believe the market is always right, this means that almost a fifth of the value of the real economy disappeared in one day. Markets can change quickly: The real economy makes measurable changes every quarter or every year, whereas financial markets can change significantly in a day. The market normally changes by less than 1 per cent on any given day, so this event was twenty times bigger than normal.

What caused such a precipitous drop? For the fundamental value of the real economy to drop this much in one day, something really big would need to happen, such as the discovery that Earth might collide with an asteroid or that we were on the brink of nuclear war. But there was no real news that day. How could such a significant financial event come out of nowhere?

Before the 1987 crash, the biggest overall drop in the stock market occurred on October 28–29, 1929, when the market lost 25 per cent in two days. The Great Depression that followed profoundly affected the entire world and lasted for more than a decade, driving unemployment in the US to 25 per cent at its peak. The 1987 crash immediately caused everyone to worry that another Great Depression was on its way, provoking mass anxiety. But, nine months later, the stock market was back where it had been, and there were few long-term economic consequences.

While debates about the causes of the 1987 crash continue, most experts believe that automatic trading by computers, called *program trading*, played an important role.[2] The following explanation is a bit technical, but it illustrates how the act of exploiting a small informational inefficiency can cause a huge allocational inefficiency. In this case the problem was the unstable feedback effects caused by the interaction of two specialized trading strategies.[3]

In the years leading up to the crash a new strategy called *portfolio insurance* was introduced, which was designed to insulate investors against a market crash. When the market goes into freefall, it can be hard to sell individual stocks at reasonable prices, whereas it is much easier to sell an S&P futures contract. (The S&P index consists of a basket of 500 large companies, and an S&P-index-futures contract allows the owner to buy the underlying stocks in the index at a given point in the future at a given price.) With portfolio insurance, when the market starts to fall, a computer automatically sells S&P futures on behalf of the owners of the insurance. The goal of the insurance is to raise cash and protect investors against losses due to sudden declines in the prices of the stocks in their portfolios.

The other trading strategy that caused problems, called *index arbitrage*, looks for discrepancies between the index's underlying stocks and the index futures. If the underlying stocks in the index are overvalued relative to the index-futures price, index arbitrageurs sell those stocks. When the stocks in the index eventually once again match the index, they buy the stocks back and make a profit.

The interaction of these two strategies caused an unintended side-effect: During the crash, the use of portfolio insurance triggered an automatic response, creating a feedback loop that couldn't be stopped, making the problem it was trying to protect investors against much worse. When the market dropped, the computers sold S&P index futures. The resulting market impact pushed the price of the S&P futures index down. Then index arbitrage kicked in. The market impact from this caused the prices of the stocks to drop even further. The combination of portfolio insurance and index arbitrage meant that the more that prices dropped, the more shares were automatically sold, which meant that prices dropped even more, and so on, throwing the market into a death spiral. Combined with the fact that investors were already nervous that the market was overvalued, this caused an enormous crash, which was only stopped when prices got so low that there were huge bargains. The brave souls who stepped in made enormous profits.

Thus we see how the unanticipated interaction between an innovation that was designed to reduce risk and a profit-making strategy that normally makes the market more efficient resulted in a gigantic departure from market efficiency: What probably should have been a small downward correction in prices turned into Armageddon.[4] Even the most sophisticated investors aren't fully rational, and almost no one anticipated this outcome. The crash offered dramatic evidence that markets can be far from efficient, and it provided a great example of systemic risk – in this case, an unanticipated (negative) emergent property of a market caused by a poorly understood financial innovation.

Even though the economy recovered quickly, the 1987 crash permanently changed views about market efficiency. Before the crash, most financial economists were, as noted, strong believers in the theory of efficient markets. But the crash forced a careful reexamination of views about markets and their relationship to the real economy, and exposed the flaws of logic based on the idea that investors are rational. If investors had been perfectly rational, they would have foreseen the dangers of portfolio insurance and behaved very differently. Market failures occur precisely because investors are *not* rational. Despite this, rational-expectations equilibrium continues to be the dominant mainstream framework for understanding markets. In contrast, in Chapters 10 and 11 we'll see how complexity economics offers a better framework for understanding why markets fail.

The market makes its own news

The October 1987 crash was just one remarkable event in a month full of them: Five of the ten largest changes in the US stock market between World War II and 1989 occurred around this time – an example of *clustered volatility*. *Volatility* refers to the amplitude with which prices vary, regardless of whether the movements are up or down. *Clustered* refers to the fact that large market moves tend to be bunched together. Most of the time, if the market is hot it's likely to remain hot, and if it's cool it's likely to remain cool.

An example is shown in Figure 10, which shows the price returns for the S&P index on successive days from 1929 to 2023. The *return* is just a name for the relative change in price – if the price increases from 100 to 105, then the return is 5 per cent. The size of the vertical

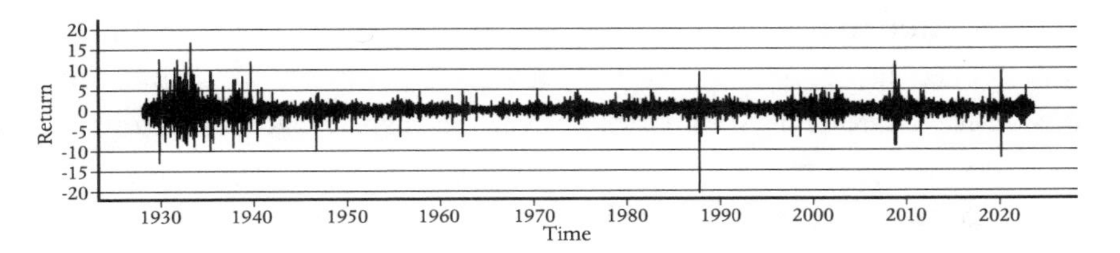

Figure 10. *The clustered volatility of price movements*. The daily returns of the S&P index measured in per cent from 1929–2023.

line is the return on a given day. The large moves are clustered together.

Without clustered volatility, large market movements like the 1987 crash would be impossible. Under Louis Bachelier's random-walk theory, if the step sizes of the drunk are all the same and he takes many steps, the distribution of possible distances he can be from the tavern after a given number of steps follows a bell curve – that is, a normal distribution. If the drunk takes n steps, each of size 1, then he can potentially be found a distance n away from the tavern, but his typical distance away is only the square root of n. For example, if $n = 10{,}000$, his typical distance away is 100 steps. Because it's extremely unlikely that all his steps are in the same direction, finding him at a distance 10,000 steps away from the tavern is so rare that for all practical purposes it never happens. Similarly, if daily stock returns followed a normal distribution with a typical size of 1 per cent, a movement of 20 per cent would be essentially impossible.

Because of clustered volatility, a more accurate model of stock prices has to take into account the drunk's changing moods. When he's feeling energetic, he takes large, rapid steps; when he's lethargic, he takes slow, small steps.[5] If he's in a particularly energetic mood, he may suddenly move a long way from the tavern. The distribution describing the size of his steps in markets has *fat tails*, meaning that large moves are far more common than they are for the *thin-tailed*, normal distribution. While really large movements like the 1987 crash are rare, they do happen a few times a century.

According to models based on rational-expectations equilibrium, any variation in the size of price returns must be due to new information, so large market moves should correspond to important new information. If stock-price movements show clustered volatility, this must be due to the clustered volatility of this new information. But, as we have already seen, there was no discernible news-related cause for the mammoth 1987 crash.

To test whether changes in stock prices are driven by news, the well-respected economists David Cutler, James Poterba and Larry Summers published a paper in 1989 entitled 'What Moves Stock Prices?'[6] Summers, who went on to become President of Harvard and Secretary of the Treasury under Clinton, had attended the 1987 'Economy as an Evolving Complex System' meeting at SFI. I've always wondered whether that conference had any influence on his 1989 paper. Summers has a remarkable pedigree as an economist, being the double nephew of two of the most famous economists who ever lived, Kenneth Arrow on his mother's side and Paul Samuelson on his father's.

Cutler, Poterba and Summers began by finding the 100 largest daily fluctuations in the S&P 500 index between 1946 and 1987. They then looked at the *New York Times* on the day after each move and recorded a summary of the paper's explanation for the price change. The authors made a subjective judgment as to whether these explanations could plausibly be considered 'real news' – or at least real enough to have triggered a sizable change in stock price. The *New York Times*'s explanations for the dozen largest daily price moves are shown in Figure 11.

The explanations for the 20 per cent drop on October 19, 1987, were 'worry over dollar decline and rate deficit' and 'fear of US not supporting dollar'. Cutler, Poterba and Summers didn't classify this as news, and I agree. 'Worry' and 'fear' are subjective statements about the emotional state of the market that have no specific reference to external events. Markets are a place where worry and fear are appropriate and normal emotions. If your investment advisor is not worried about losing your money, you should fire her.

Does news explain clustered volatility?

Largest S&P index moves 1946–87

(Cutler, Poterba, Summers 1989)

Rank	Date	%	New York Times explanation
1	October 19, 1987	–20.5	Worry over dollar decline and rate deficit Fear of US not supporting dollar
2	October 21, 1987	9.1	Interest rates continue to fall Deficit talks in Washington Bargain hunting
3	October 26, 1987	–8.3	Fear of budget deficits Margins calls Reaction to falling stocks
4	September 3, 1946	–6.7	'No basic reason for the assault on prices'
5	May 28, 1962	–6.7	Kennedy forces rollback of steel price hike
6	September 26, 1955	–6.6	Eisenhower suffers heart attack
7	June 26, 1950	–5.4	Outbreak of Korean War
8	October 20, 1987	5.3	Investors looking for quality stocks
9	September 9, 1946	5.2	Labor unrest in maritime and trucking
10	October 16, 1987	–5.2	Fear of trade deficit Fear of higher interest rates Tension with Iran
11	May 27, 1970	5.0	Rumours of change in economic policy 'Stock surge happened for no fundamental reasons'
12	September 11, 1986	–4.8	Foreign governments refuse to lower interest rates Crackdown on triple witching announced

Figure 11. *Does news explain clustered volatility?* The 12 largest moves in the S&P index from 1946 to 1987. The first column ranks the moves, the second gives the date, the third the size of the move, and the fourth a summary of the *New York Times* explanation. Moves that Cutler, Poterba, and Summers rated as 'real news' are shown in grey.

The first example of real news among their list is the fifth-largest price movement, on May 28, 1962; the news that day was 'Kennedy forces rollback of steel-price hike'. This is something that happened outside the market which should have affected investor decisions. Other examples of significant news items are 'Eisenhower suffers heart attack' and 'outbreak of Korean War'. Of the dozen largest price fluctuations, only four were attributed to real news events, a ratio that they found also roughly applied to the largest 100 moves.

Their paper offers strong evidence that the market frequently makes its own news – that is, it often operates according to its own internal dynamics rather than responding to external events. The clustered volatility of the market is also evidence of this phenomenon: The three largest moves between 1946 and 1987, and five of the twelve largest moves during those years, all occurred during October 1987. The market was extremely volatile that month, but none of the fluctuations were driven by external news. You might also notice that nine of the twelve biggest movements were negative. Prices go down faster than they go up!

Some of the explanations for these biggest fluctuations stand out. The reason given by the *New York Times* for the fourth largest move, on September 3, 1946, was 'no basic reason for the assault on prices'. I doubt that explanation will ever appear in the *New York Times* again, even if it's true; the press always gives a reason, whether or not it's justified. Perhaps journalists were more honest in the old days? Another good explanation was 'investors looking for quality stocks'. Did they mean to imply that on other days investors were looking for inferior stocks?

Cutler, Poterba and Summers's conclusions have since been further substantiated.[7] Subsequent studies refute the efficient-markets hypothesis (at least in terms of allocational efficiency). While news events can influence stock prices, the majority of price movements have little to do with current affairs. Markets appear to have their own internal dynamics.

Can we measure financial-market efficiency in quantitative terms? This question was put to a panel of experts at a three-day conference that John G. and I organized at SFI in May of 2000, called 'Beyond Equilibrium and Efficiency'. We invited a group of leading financial economists – including leading behavioral economists, such as Robert Shiller and Richard Thaler – along with leading econophysicists and complexity economists. The mainstream economists spoke on the first day, the complexity economists and econophysicists spoke on the second day, and the third day was for discussions. This proved to be a naïve way to organize the conference: Almost all the mainstream financial economists skipped the second day and went sightseeing. As a result, the discussion on the third day was not very enlightening.

Nonetheless, we did have a lively formal debate about market efficiency on the evening of the conference dinner. There were two teams, with three people arguing in favor of market efficiency and three arguing against. The latter team consisted of Shiller, Sanford (Sandy) Grossman and me. As measured by the crowd reaction among the mainstream figures, we lost the debate. It wasn't even close. This was in part because my two teammates largely agreed with the other side. Somehow, during the discussion, a consensus emerged among the mainstream economists that 'markets are about 97 per cent efficient'. As they would freely admit, we don't have a good way to measure deviations from efficiency in quantitative terms. Looking back, the debate was confused by the fact that we weren't clear about what this really meant.

In hindsight, I would answer that the efficiency of the market depends on which kind of efficiency you are talking about. It is indeed hard to beat the market. The fact that the signal-to-noise ratio of strategies like Prediction Company's are about 1:20 indicates that markets are close to informational efficiency. But the Cutler, Poterba and Summers result indicates that only about a third of daily market movements occur in response to meaningful external events. This suggests that, from an allocative point of view, the market is about 33 per cent efficient. My favorite summary of this dichotomy is due to Fischer Black, one of the inventors of options pricing. In a paper titled simply 'Noise', he made a tongue-in-cheek statement that I will paraphrase as, 'I believe in efficient markets: prices are within a factor of two of fundamental values 90 per cent of the time.'[8] By this he meant that, yes, prices loosely reflect fundamental values, but they often fail to do so very well. This was borne out in a 1981 paper by Robert Shiller, who compared the Dow Jones index to the valuation of the companies in the index based on their dividends.[9] The correct way to compute fundamental values is debated, but Shiller's study suggests that prices make large swings from fundamental values, indicating that markets are on average less than 50 per cent efficient. (Which is why I was particularly disappointed when he agreed with the 97 per cent argument.)

This is not just an esoteric debate. As we will see in the next chapter, the fact that markets can stray far from fundamental prices on their own means they can malfunction, misguiding the economy. As the crisis of 2008 showed, this can have serious consequences, like causing unemployment to surge. Acknowledging that markets are inefficient could lead us to theories that would help us understand what causes malfunctions so that we can better guide our economy.

While the Cutler, Poterba and Summers paper is widely cited, its main lesson seems not to have been fully absorbed by mainstream economists. To my knowledge, no extant equilibrium model offers a reasonable explanation for clustered volatility that is consistent with their results. The only models that plausibly explain it – those we are about to discuss – allow the market to be out of equilibrium.

Complexity economics captures the moods of markets

The Cutler, Poterba and Summers study suggests that it is 'as if' the market has moods. What causes them? Several of the early complexity-economics models explain this 'moodiness'.

Brian Arthur and John Holland were among those in attendance at the 1987 SFI meeting. Though never really part of the mainstream, Arthur was an economist at Stanford at the time. Holland was a computer scientist and one of the founding figures of complex-systems science and machine learning, whom I had come to know well after inviting him to visit Los Alamos in 1982. After the 1987 meeting, the two of them teamed up with Blake LeBaron, an economist who was then a graduate student of José Scheinkman and now teaches at Brandeis University, and the physicist Richard Palmer. These four reconvened at SFI in subsequent years and formed an interdisciplinary research team seeking ways to explain clustered volatility in financial markets. They built an agent-based model for financial markets now called the Santa Fe Artificial Stock Market model.[10] Their goal was to understand what causes the market to stray from equilibrium and make its own news. To do this, they built a computer simulation of a market with heterogeneous agents using diverse investment strategies.

The model's agents were simulated traders, who were offered a choice between investing in a stock or a bond. The stock paid dividends of variable size, while the bond paid a steady (but lower) interest rate. Each simulated trader was a boundedly rational agent with an 'artificial brain' based on a machine-learning algorithm, invented by John Holland, called a *classifier system*.[11] Their artificial brains allowed them to form their own expectations and make decisions based on past experience. The agents could choose their source of information. They could take account of the recent behavior of prices, and base their decisions on these and other technical indicators, or they could base their decisions on fundamentals, such as dividend payments, or they could do both. Unlike a real market, which is full of unknowns, in the Santa Fe Artificial Stock Market the researchers were like gods: They knew everything about their artificial world because they had created it themselves. They knew the statistical process by which the stock paid dividends. This allowed them to experiment with imbuing all the simulated traders with perfect rationality by overriding the classifier system and replacing it with expectations that were objectively correct within their artificial world. When they did this, as expected, the market obeyed the rational-expectations equilibrium: The stock price stayed near the correct valuation based on its dividend stream and fluctuated only when new information arrived. In this experiment, the market was perfectly efficient and everything happened just as rational expectations predicts it should.

However, when the agents in the model lost their divine guidance and became boundedly rational, forming their own views with the classifier system, the market's behavior became more realistic. In this case, current and future prices remained largely uncorrelated, as they are in the real world (and as predicted by rational expectations). However, the volatility of prices was much higher than predicted by rational expectations (as it is in the real world). Furthermore, the volatility was clustered – quiet periods were interspersed with bursts of noisy behavior, as happens in a real market. The market appeared to be making its own news, just as Cutler, Poterba and Summers's study says the real market does. The clustered volatility of the Santa Fe Artificial Stock Market model was an emergent property, generated by the interactions of the market with itself.

To figure out what caused the clustered volatility, the researchers could look inside the brains of their artificial traders. They found that the artificial traders spontaneously formed an ecosystem of specialists. Some of them became value investors, focusing their attention on understanding the dividend stream. Others became technical traders, dedicated to understanding patterns in prices. Many of these technical traders were trend followers, who bought the stock whenever its price started to trend upward and sold it when it trended downward. The relative proportions of value investors and technical traders fluctuated: When the trend followers became active, the market was volatile; when they were inactive, it was quieter. Individual artificial traders sometimes switched between these two strategies, but most of them chose a strategy and stuck with it.

The Santa Fe Artificial Stock Market was a complicated model, requiring serious software effort to reproduce. In parallel research, published around the same time, William 'Buz' Brock of the University of Wisconsin and the Dutch mathematician/economist Cars Hommes found similar results using a much simpler model.[12] Their model's set-up was similar to that of the Santa Fe Artificial Stock Market model; the agents chose between a stock and a bond. But in the Brock-Hommes model, agents were given an explicit choice between a value-investor rule and a few simple technical trading rules. They could monitor the success of each rule and switch to the ones that recently performed best. At different times, different rules tended to dominate. Brock and Hommes observed clustered volatility and fat tails in price returns, just as in the Santa Fe Artificial Stock Market model.[13] Since these models came out, many other agent-based models have explored similar ideas and confirmed the same basic results. The best models achieve a good quantitative match to real data, generating uncorrelated directional price movements, clustered volatility and price-movement distributions that closely match real markets.

The underlying explanation of this behavior is essentially ecological. During calm periods, the fundamentalists dominate. However, if for some reason prices happen to move in the same direction several times, trend followers start to do well, more of the agents are attracted to trend following and their influence on the market grows. This generates a positive feedback loop, launching a boom in which prices exceed fundamental values, which causes the value investors to sell. When the trend followers eventually run out of new capital, the upward trend falters, and, when prices drop, the trend followers start selling too, and prices converge toward fundamental values again.

These models provide a plausible explanation for the lack of correspondence between price movements and news. In Chapter 11, we will explore another family of agent-based models – those that study the implications of the use of credit, which also leads to clustered volatility. Both families of models assume bounded rationality and allow deviations from equilibrium.

There are two reasons that agent-based models for markets haven't received much attention from mainstream economics. The first is that, like any agent-based model, they run counter to the central tenets described in Chapter 5. Another, better reason is the lack of useful quantitative predictions. These models provide an explanation for *why* clustered volatility occurs, and they do a good job of matching the statistical properties of real markets, but so far they have no useful practical applications.

This is changing: For example, it was recently shown that models of this type can provide better predictions of clustered volatility than standard models.[14] They are also beginning to be used by banks to understand and even predict systemic market failures, and I am hearing more and more rumors that some hedge funds are using them. We'll discuss these practical developments at the end of Chapter 11, but, before we do so, it's worth further developing the conceptual framework provided by my theory of market ecology, which helps explain how and why markets deviate from efficiency, and how this can cause them to malfunction.

10.
Ecology and Evolution in Finance

My journey from market anthropologist to market ecologist

When Prediction Company began, in 1991, our expertise was purely in making predictions based on historical data. We needed funding and none of us knew anything about financial markets, so we decided to look for a partner with the capital and financial expertise we lacked. Someone needed to go to Wall Street, and I was the chosen delegate. Since I had never owned a suit, everyone felt that the company should buy one for me, but in the end I paid for it myself. (After describing my mission to the fussy sales assistant in the best men's shop in Santa Fe, he spent hours making sure I had exactly the right cut, the right tie and the right kind of shirts, shoes and socks.) I began making trips from Santa Fe to New York, Chicago and San Francisco, searching for a partner. In one year, I called on at least twenty Wall Street firms, including Kidder-Peabody, Salomon Brothers, Goldman Sachs, Renaissance, D. E. Shaw, Tudor Investment, Citibank, Bank of America, and O'Connor and Associates. I was struck by how different these firms were from one another – a difference reflected in their trading strategies, their vocabulary and even the way they dressed and talked.

I felt like an anthropologist studying another culture. In trying to learn their jargon and worldviews, I quickly realized that they belonged not to a single tribe but to different tribes within a larger community. Even in the same firm, there were sometimes distinct subcultures: At Bank of America, on one floor everyone told me that fundamentals were all that mattered, while on the floor below everyone said fundamentals were useless and technical trading signals were the only thing worth paying attention to. I encountered many different strategies for making money. To my surprise, most traders were happy to talk about them, at least at a high level.

I came to regard finance practitioners as skilled specialists with detailed domain knowledge, who were nonetheless far from rational in the way economists use this term. A perfectly rational agent would be able to exercise all possible strategies at once, whereas real traders are boundedly rational, typically pursuing a single strategy, or a few at most. Standard theoretical concepts like equilibrium and efficiency seemed of little use for understanding what traders did, why they did it and how this affected markets.

I turned instead to biology. The fact that investors are specialists makes ecology the natural framework. Ecology provides a useful metaphor for understanding specialized behavior in the economy in general, but for financial markets I realized that it could be made mathematically precise and used to explain many of the phenomena I observed. This provides a deeper, more realistic understanding of how markets work and why they sometimes don't work. While I think ecology is a useful explanatory principle for all of economics, financial markets provide a forum where it is easy to make the ideas crisp and quantitative.

Let's start with the big picture. To understand why ecosystems are the way they are, and how they change through time, requires a theory of evolution. The financial ecosystem satisfies the essential elements of the theory of evolution, which Darwin called 'descent with variation and selection'. Here, *descent* means continuity across generations, *variation* means offspring differ in some ways from their parents, and *selection* means that only successful individuals survive and reproduce. Darwin and Wallace showed that these three elements explain how new species emerge and, over time, change the biosphere.

The same three essential elements characterize the agents and their strategies in the financial system. There is descent because investors develop their strategies by inheriting knowledge from their mentors, as well as learning from their own prior experience. There is variation because investors tinker with their strategies to improve them. There is selection because strategies compete for profits; successful strategies accumulate wealth and unsuccessful strategies go extinct.

Of course, there are important differences in the way organisms and financial strategies evolve. Biological organisms interact by breeding, eating one another, and reshaping their common environment, whereas trading strategies interact via the market, reshaping patterns in the way prices are formed as they eliminate and create market inefficiencies. The distinctions among trading strategies are fuzzier than those among biological species. Unlike biological organisms, which reproduce following the fixed rules of biochemistry, financial strategies have no genetic code and instead propagate via cultural transmission. People design strategies, so their continual variation isn't random; this accelerates the pace of evolution. Selection in biology affects the population size of each species, whereas in finance it affects the capital invested in each strategy, which I will call its *wealth*.

Whether or not competition plays a fundamentally different role in markets than it does in biology is a subject of debate. Competition is a core assumption in financial economics, but mainstream economists reach different conclusions about its consequences than biologists do. Economists like Milton Friedman, who was Eugene Fama's thesis advisor, used it to shore up the efficient-markets hypothesis, arguing that competition should quickly eliminate market inefficiencies. In contrast, Darwin and Wallace used competition to explain why organisms specialize and how this creates new species. Mainstream economists argue that competition leads to a state of rest, whereas evolutionary biologists see it as the driver of perpetual change.

Why did economists and biologists arrive at opposite views about the role of competition? The key difference lies in their assumptions about the reasoning power of agents. As noted, when Fama and Friedman postulated efficient markets they also postulated rational agents – perfect problem-solvers. If such rational agents really existed, the market would always be calm and stable – 100 per cent efficient. But biological organisms are agents who follow hardwired strategies that evolve through time to improve their reproductive rates, in an environment that is perpetually subject to change from both external and internal influences. Trees or bacteria behave in a hardwired manner, whereas rational agents can solve any problem and alter their strategies instantaneously.

People can indeed think, but their finite cognitive capacity places limits on their successful strategizing. As discussed in Chapter 6, perfect rationality provides a useful model for understanding human behavior when the problems are sufficiently simple – options pricing or tic-tac-toe are examples. But to explain market malfunctions, which generally occur in complicated and unforeseen situations, specialization and bounded rationality matter, and the basic principles of evolution and ecology provide a better starting point than the efficient-markets hypothesis.

The proliferation of computerized trading strategies, like ours at Prediction Company, underscores the bounded rationality of agents in the market. While our strategies were constructed through a careful and scientific search carried out by intelligent people, the final code that was implemented to trade our strategies was very simple. It consisted of classic stimulus/response instructions, of the form 'If conditions A, B and C are met, then buy (or sell) the following assets.' Ants are far more intelligent than our code was. It's difficult to estimate precisely how much stock trading is now done by computers, but most estimates put it well in excess of 50 per cent. Computerized trading is far from rational, but it's used in many domains because it's more successful than the trading done by human beings, not to mention cheaper and quicker. If we were fully rational, we would have foreseen the problems that program trading or mortgage-backed securities would cause.[1]

The theory of market ecology

My musings about market anthropology and our experience measuring market impact led me to a theory of market ecology.[2] In my spare time at Prediction Company I began scribbling and solving equations, which eventually led to a paper titled 'Market Force, Ecology and Evolution', which I finished and posted online in 1998. In a nutshell, the basic ideas are that:

- Financial traders are boundedly rational, so they use specialized strategies. Their strategies can be sorted into groups, like species in biology, even if the boundary between groups can be fuzzy.
- Financial strategies feed off inefficiencies and interact through prices. Trading has market impact, which changes prices and affects the profits of each strategy.
- As a result, the wealth of each strategy evolves through time. As the wealth invested in a strategy becomes larger, it has a bigger effect on the market.
- This causes market inefficiencies to evolve through time. While markets may tend toward an efficient state, the path there is bumpy, the journey is slow and they never really reach their destination.
- If a destabilizing strategy becomes too dominant, it can unsettle the market and cause it to malfunction. To understand why markets malfunction we need to classify financial strategies, understand how they affect the market and track their evolution through time.

Let's now explore this in more depth.

The diverse anthropology of traders reflects the diversity of their specialized strategies, which have names like trend following, technical trading, value investing, index arbitrage, market making, derivative trading and so on. Small firms typically only do one of these; big firms may do many, but with an aggregation of specialized groups, who may have little interaction with each other.

To a trader, the word *inefficiency* is just a fancy name for a pattern in prices that can be exploited to make profits. The ultimate origin of inefficiencies is economic activity, such as raising capital and offloading risks, but, just as in biological ecosystems, strategies also exploit the inefficiencies created by other strategies. Financial strategies take advantage of these inefficiencies, but, in doing so, alter the market and create new inefficiencies, and in some cases cause market malfunctions.

For example, a feature of economic activity that creates inefficiencies is its persistence. Successful companies have momentum, meaning if they are growing rapidly they are likely to continue growing rapidly, at least for a while. This makes price changes persistent, creating a niche for trend-following strategies, which buy when markets are rising and sell when markets are falling. This can be profitable, but it can also drive prices away from fundamental values. Trend strategies are destabilizing, so much so that if there are only trend strategies in the market, prices blow up (head for infinity).[3] Trend strategies can play a major role in fueling bubbles, such as the dot-com bubble which peaked in early 2000. Stock prices in the Nasdaq index went from under 1,000 in 1995 to a peak of 5,048 on March 10, 2000, and then dropped to 1,114 on October 9, 2002. The huge swings in stock market value had important economic consequences: At the peak of the bubble, it was easy to get almost any new, or not-so-new, internet-related idea funded. After the bubble popped, it became difficult to get even really good ideas funded, and the pensions of many people who had the misfortune to invest in tech stocks just before the bubble popped were demolished. This is a great example of how imbalances in the market ecology – in this case overpopulation of destabilizing strategies – can make markets malfunction.

Eugene Fama initially resisted the idea that trend following could be profitable, because it seemed irrational, but he and his collaborator Kenneth French eventually looked at the data and realized that it could be surprisingly profitable.[4] This contradicted efficient-markets theory, but they fixed this by moving the goalposts, incorporating such profits into their benchmark for efficiency instead of calling them *excess profits*.

Strategies affect each other's profits via their market impact on prices, and thinking in ecological terms can help understand their interaction. For example, trend strategies can range from short-term strategies that buy when prices go up for a short period of time (say a week) to those that only react after much longer periods of time (say a year). Suppose the market happens to have a good week, for whatever reason. As a result, the short-term trend strategies buy. This drives prices up further, causing slightly longer-term trend strategies to buy, causing prices to go up more, and so on, until finally after a year the price has risen substantially and all the trend followers have bought the asset. The price will eventually drop, generating a wave of selling as all the trend followers unwind their positions.

This self-fulfilling dynamic is good for the short-term trend strategies but bad for long-term trend strategies (whose profits are diminished by the presence of the short-term strategies). To put this in ecological terms, short-term strategies have a predator–prey relationship with long-term strategies. An ecological analog might be rabbits and foxes: An increase in the rabbit population causes the fox population to grow, and an increase in the fox population causes the number of rabbits to drop. Similarly, an increase in the wealth of long-term trend followers causes the returns of short-term trend followers to grow, but an increase in the wealth of short-term trend followers causes the returns of long-term trend followers to drop.

The possible relationships between trading strategies can be classified using biological nomenclature, by applying 'what-if' experiments. What if the wealth of strategy A increases? How does that affect the returns of strategy B? Similarly, if the wealth of strategy B increases, how does this affect the returns of strategy A? If both of their returns decrease when the wealth of the other strategy increases, then they have a competitive relationship. If A's returns go up when B's wealth goes up, but B's returns go down when A's wealth goes up, then they have a predator–prey relationship, in which A is the predator and B is the prey. If both of their returns increase when the other's wealth goes up, then they have a mutualistic relationship. An important case is when A = B. The wealth of a strategy strongly affects its own returns as well: The laws of supply and demand imply that, absent market manipulation, the returns of a strategy will decrease as its wealth increases, so strategies automatically compete with themselves.[5] Sometimes two strategies simply don't interact, so that a change in the wealth of strategy A has no effect on the returns of strategy B. But because trading affects prices, and almost all strategies are price-dependent, many if not most strategies interact.

In my market-ecology paper, these ideas allowed me to derive a set of equations describing how the wealth of strategies changes based on their ecological interactions. My equations are similar to the most famous equations in ecology – the Lotka-Volterra equations, named in honor of the American mathematician Alfred J. Lotka and the Italian mathematician Vito Volterra, who first wrote them down in the early twentieth century.[6] The Lotka-Volterra equations showed that predator–prey systems could behave in a surprising way. Independently, they showed that under certain circumstances the populations of rabbits and foxes could oscillate indefinitely. The equations are only an approximation – nothing ever happens truly indefinitely – but there are many examples in real ecosystems where populations of predators and prey markedly fluctuate over long periods of time, without any obvious external stimulus. (In fact, more realistic generalizations of the Lotka-Volterra equations often display chaotic oscillations.)

My equations are a general version of the Lotka-Volterra equations, but with the population of species replaced by the wealth of strategies. As in their equations, there are some situations where the wealth of all the strategies settles down to a constant value, approaching a fixed-point attractor, and other situations where they oscillate indefinitely. So, for example, the population of trend followers may go up while the population of value investors goes down, and then the opposite happens. So far we lack data to properly fit these equations to financial markets, and my derivation relies on approximations that may not always be valid. At this point, it remains to be seen whether sustained oscillations ever occur in practice, but the theory makes the correspondence to biological populations tangible.

I submitted 'Market Force, Ecology and Evolution' to one of the more avant-garde economics journals, but the referees didn't really understand it and, in the end, they wanted to publish only the first half, dropping the discussion of evolution and much of the ecology. The resulting published paper, which was co-authored with my research assistant Shareen Joshi, gave only part of the story.[7] I despaired of ever publishing the full paper until several years later I happened to meet Giovanni Dosi, a radical Italian economist who has developed and promoted evolutionary ideas in economics, and who is the head of one of the world's largest complexity-economics groups, based in Pisa, Italy.[8] Despite working well outside of mainstream economics, Giovanni was then editor-in-chief of the Oxford journal *Industrial and Corporate Change*; he liked my paper and published it there. It caught the attention of Andrew Lo of MIT, a prominent financial economist and an original thinker who manages to be a member of the mainstream while also advocating ideas that are off the beaten path. Like me, he felt that economics could benefit from borrowing more ideas from biology, and our discussions resulted in a joint paper called 'Frontiers of Finance: Evolution and Efficient Markets', published in 1999 in the *Proceedings of the National Academy of Sciences.*[9] Andrew went on to develop what he calls the *adaptive-markets hypothesis*, which embraces market ecology and evolution but also introduces other interesting ideas from biology, artificial intelligence and the cognitive neurosciences.[10]

While 'Market Force, Ecology and Evolution' attracted interest from econophysicists, complexity economists and a few open-minded mainstream economists like Andy, it was otherwise ignored. I realized that the only way to make mainstream economists pay attention to theories that run counter to the accepted canon is by showing that they lead to superior empirical predictions – and, even then, the evidence needs to be so overwhelming that it cannot be ignored.

Can market ecology predict market malfunction?

In order to study market ecosystems empirically, I needed access to special data. There is plenty of data available about market transactions, but it lacks *counterparty identifiers* – a record of who transacted with whom. Without counterparty identifiers, I was like an ecologist whose data say 'some animal ate another animal', without naming the species involved. Most regulators have this kind of data, but, despite several attempts, I never managed to get any of them to share it with me. Without it, I couldn't see how to empirically show that the underlying ideas of market ecology were useful. Since I couldn't find the data I needed, I suspended my research on market ecology.

Then, twenty years later, in 2018, I attended a conference organized by Andy Lo and Simon Levin on applications of concepts from biology to economics.[11] Simon is a professor of ecology at Princeton, and a fellow member of SFI's external faculty, who has collaborated with Andy on some of his ideas about adaptive markets. The meeting brought together a diverse group of interdisciplinary thinkers, including economists who thought in biological terms and biologists who were interested in economics. This rekindled my interest, spurring me to do some new work – an opportunity that fortunately coincided with the interests of my Dutch graduate student, Maarten Scholl. Maarten had been admitted into the Computer Science Department at Oxford, and he was looking for a project. I realized that my earlier paper had left many theoretical questions unanswered, so Maarten and I began working on market ecology, along with his other thesis advisor, Ani Calinescu.

Maarten built a 'laboratory' consisting of a computer simulation of an idealized stock market. The set-up was similar to the Santa Fe Artificial Stock Market, but it was simpler, and we had a different goal. Our purpose was not to make a realistic simulation of a market but rather to use the theoretical framework of ecology to better understand how deviations from market efficiency evolve through time, in a transparent setting where we could see exactly what caused what. In our market, there was a single stock paying a randomly varying dividend and a bond paying a fixed interest rate. There were three trading strategies – a value investor, a trend follower, and a 'noise trader' who represented a nonprofessional investor making mostly random decisions. Each of the strategies was initially endowed with wealth, consisting of shares of the stock and the bond; as they traded them through time, their wealth evolved in proportion to their profits and losses. While the wealths of the three strategies tended to adjust so that the market became somewhat more efficient, it did so slowly and with considerable uncertainty. Even when we waited for simulated centuries, there were still substantial fluctuations in the wealths of the strategies, corresponding to deviations from perfect efficiency.

We used our laboratory to show explicitly how it is possible to apply concepts from ecology to understand markets.[12] We classified the pairwise relationships between the strategies as competitive, predator–prey and mutualistic, and showed how these relationships could change with the changing wealth of the strategies.[13] As expected, the trend follower and the value investor tended to prey on the noise trader, but this was not always true, and we even observed mutualistic relationships.[14] We mapped out the 'food web' of our financial system, demonstrating how the strategies utilize inefficiencies to maintain their wealth, and how they create inefficiencies that support one another.

We wanted to demonstrate how the theory of market ecology can predict market malfunctions. By *malfunctions*, we meant deviations of the prices of assets from their true value or volatility of prices in excess of what should occur based on external information (as discussed in Chapter 9). Because trend-following strategies are inherently destabilizing, we expected that when that strategy has more wealth than the other two, we would see larger mispricings and more volatility. This intuition was borne out: The changes in the wealth of the strategies were highly predictive of both mispricings and volatility.

But is there solid, empirical evidence that market ecology is useful? Over the last two years, with a little help from Maarten, my persistent and creative French mathematics student Aymeric Vie has developed an empirically-grounded simulation of the US stock market that demonstrates that the answer to this question is 'yes!'. The simulation makes use of regulatory filings by mutual funds, who are required to submit their positions to the Securities and Exchange Commission (SEC) every quarter. This makes it possible to back-out crude approximations of the strategies of long-term investors, who hold their positions for more than three months. Aymeric explicitly simulates the trading of thousands of mutual funds: Every quarter his simulation is initialized with the actual holdings of the funds. It then simulates their trading and predicts how it will affect next quarter's prices. He has shown that his simulation has predictive power above and beyond fundamentals alone. (*Fundamentals* refers to information that affects the price for economic reasons, such as earnings and dividends.) This shows that the market is not perfectly efficient: The type of funds that are active at any point in time affects prices. Market ecology has predictive power.

Regulatory implications of market ecology

Maarten and Aymeric's work provides a proof of principle for a new, more effective way to regulate markets. Regulators such as the SEC, who have access to much better data, could use our methods to simulate markets, monitor their stability and intervene when they are in danger of serious disruption. Such a simulation could also provide a testing ground for new financial instruments.

One of the challenges in ecology is understanding what happens to an ecosystem when a new species is introduced or an existing species is removed. For example, the widespread hunting of beavers for their pelts in America in the nineteenth century had a host of negative side-effects. Beavers eat young trees and thereby create meadows; their ponds raise the water table and provide essential habitat for many plants and animals. Another famous case is the introduction of rabbits into Australia in 1859. Rabbits have no natural predators in Australia, and, by 1920, there were roughly 10 billion of them.[15] Rabbits eat tree seedlings, denude vegetation and have contributed to the loss of a vast number of native Australian species. To control their population, other non-native species, such as foxes and cats, were introduced, but these preyed on other small animals as well and have caused even more environmental damage.

In the United States, major projects require environmental-impact statements that analyze ecological effects. But there is no equivalent impact statement for a new type of derivative or financial exchange – at least in part because the belief in efficient markets has suppressed development of the science necessary to do this. There are many clear instances where changes in market structure or the introduction of new financial instruments caused serious problems. As noted, the introduction of portfolio insurance fueled the 1987 stock market crash, and the introduction of mortgage-backed securities was a primary cause of the financial crisis of 2008. To understand the likely effect of future financial innovations, regulators could make old data with counterparty identifiers available to researchers, who could then develop quantitative agent-based models of financial markets and use them to test market innovations for possible adverse side-effects. We could also monitor the financial ecosystem in real time to anticipate problems and intervene when necessary.

An agent-based model built for the Nasdaq stock exchange is proof that this can work.[16] Prior to 1997, US stock prices were traded in increments of an eighth of a dollar. Then, American stock exchanges were ordered by the SEC to begin trading in increments of a penny, a process called *decimalization*. To try to understand any possible adverse side-effects, Nasdaq hired the BiosGroup, a consulting company of complex-systems scientists founded by Stuart Kauffman. In an effort led by the complex-systems scientist Vince Darley, the BiosGroup built an agent-based model for trading on the

Nasdaq. The agents in the model were traders who followed specific strategies. The role of market-makers was particularly important. The model made six predictions, five of which were borne out when the change to decimalization was actually made. (For example, it correctly predicted that spreads would narrow, and that the depth of the market – that is, the size of the trades it could absorb without affecting prices – would drop as well.)

Several regulatory institutions, such as the Bank of England and the European Central Bank, are beginning to collect the necessary detailed data, and to monitor the balance sheets of major banks and other important financial institutions. Some of them are also beginning to build agent-based models that can use this data, though none of these efforts is properly funded. We'll say more about this in the next chapter when we discuss the destabilizing effects of leverage.

The market mass-extinction of August 2007

Convergent evolution is another concept from ecology that is useful for understanding markets. There are many species which, despite having very different ancestry to one another, evolve to inhabit the same niche and perform similar functions. For example, hummingbirds and hawk moths both hover in the air with their wings beating rapidly while extracting nectar from flowers. They can be hard to tell apart from a distance, but up close they look completely different, which is not surprising given that their last common ancestor lived hundreds of millions of years ago.

Convergent evolution also happens in financial markets. Even without any contact, people with similar training often find similar solutions to the same problems. Prediction Company was what is known as a statistical arbitrage (or stat-arb) fund. As a result of convergent evolution, Prediction Company's performance in the late 1990s became closely correlated with that of other stat-arb funds; our niche became more crowded. As in biological overpopulation, this was inimical to survival. To illustrate the effects of overcrowding: One of our main competitors was an investment-management firm called D. E. Shaw, which allied with Bank of America in 1997. An unrelated problem at Bank of America forced D. E. Shaw to temporarily shut down its stat-arb fund. In the nine months or so that D. E. Shaw halted its trading, we made profits at a rate much higher than normal. As soon as they began trading again, this stopped. This was because, in ecological terms, we were competitors, occupying the same niche; our trades were similar to theirs, and we had common market impact, which lowered both of our returns.

While there were only a handful of stat-arb funds in the mid-1990s, by 2007 there were hundreds, most of them trading as much capital as they possibly could. In August 2007, after I had left Prediction Company, I got a call from Karen Lawrence, CEO of Prediction Company and a close friend. She asked if I had heard about anything unusual going on in the markets. Prediction Company was having a strange problem: It had taken substantial losses for several days in a row, even though the market was otherwise normal and there was nothing in the news. Almost all of the stocks Prediction Company owned went down, and almost all of the stocks it had short positions in went up. Though none of the individual losses were large, Prediction Company was taking systematic losses on almost all of its 1,000 or so positions. This had never happened to statistical arbitrage firms before, even if there were antecedents in other markets.[17] Company researchers concluded that the only possible cause was that other stat-arb funds were liquidating their positions.

It turned out that Prediction Company was right. The trigger was a prominent investment bank with large positions in mortgage-backed securities. As bad news emerged about housing markets, the investment bank needed to raise cash to cover its anticipated losses. To do so, the bank's managers decided to liquidate the positions of their large stat-arb fund in order to free up capital for the rest of the bank. To reduce market impact, such liquidations are spread out over time, and can take weeks. The persistent pressure on prices caused other stat-arb funds, who all had similar positions, to take losses, and, once these were sufficiently large, they were forced to liquidate as well. This set off what came to be known as the 'quant meltdown', in which the market impact of liquidations systematically pushed prices in the wrong direction. Prediction Company was confident that, as soon as everyone stopped liquidating, profits would bounce back – and so they did. The profits of the few survivors soared.[18]

There were several surprising things about the quant meltdown. First, it was not obvious beforehand that a niche could get so crowded that the simple act of a large player liquidating its position could bring the whole sector down, causing a mass extinction.[19] We now understand that this is partly a consequence of the fact that stat-arb funds use leverage, which we will examine in Chapter 11. Second, the trigger that brought them down came from mortgage-backed securities, which are unrelated to any of the assets traded by stat-arb funds. But the fact that several large investment banks with stat-arb funds also held mortgage-backed securities created a link between the two.

The quant meltdown was a disaster, but at least it was self-contained. If you weren't investing in stat-arb funds, you likely didn't even notice it. Nonetheless, it was a harbinger of far worse things to come. Less than a year later, the link between mortgage-backed securities and lending in general severely stressed the world's financial system and caused a major recession. If the regulators had been monitoring markets by running agent-based models that could warn of systemic risk, they might have acted sooner to avert or dampen the financial crisis of 2008. We'll investigate this further in the next chapter.

11.

How Credit Causes Financial Turbulence

The most recent global financial crisis reminded the current generation of the lessons that their grandparents had learned in the Great Depression: The self-regulating economy does not always work as well as its proponents would like us to believe.

Joseph Stiglitz[1]

Credit, leverage and financial crises

Providing credit, including loans, mortgages, bonds and credit cards, is one of the most important functions of the financial system. The details of contracts for offering credit vary, but the basic mechanism is the same: A lender temporarily provides money to a borrower. Credit is essential for the operation of the economy – people need it to buy houses, businesses need it to expand their operations – but it also introduces risks.

Credit connects us together by linking the balance sheets of borrowers and lenders. Contracts for the provision of credit form a vast network, with balance sheets as nodes, and links from lender to borrower. Borrowers often use more than one source of credit, lenders supply money to multiple borrowers, and the same agent can simultaneously both lend and borrow. The network of credit contracts is an important subset of the web of contracts. The credit network is also closely coupled with the rest of the economy. If the economy goes into a depression and you lose your job, you may be unable to repay your loans; if businesses can't get credit, GDP suffers. Upheavals in the credit network are the main cause of recessions.[2]

The overextension of credit was a major cause of the 2008 financial crisis.[3] When credit is used to buy assets, it's called *leverage*. Buying assets with borrowed money may sound imprudent, but people do it all the time. Suppose you want to buy a home for $500,000 but you can't pay the full price in cash. You take out a mortgage, putting 20 per cent down and borrowing $400,000 from the bank. Leverage is defined as the ratio of assets to equity. Thus, your leverage is now 5:1, since the value of the house (the asset) is five times the value of your equity (the 20 per cent down payment).[4]

The term *leverage* comes from a mechanical analogy to a lever. Just as a lever lets you apply a larger force than your limited strength allows, borrowing lets you buy more than your current assets permit. Most of us couldn't buy a home without leverage. It is essential for businesses, which often need more resources than they can marshal with their own capital. But unlike mechanical leverage, which amplifies force in a straightforward way, financial leverage does more than just amplify profits and losses – it creates risk and can destabilize markets.

Leverage can be risky for both borrower and lender. Suppose that immediately after you buy your house for $500,000, its value drops by 10 per cent. You still owe the bank $400,000, but your house is now worth only $450,000. Your equity in the house is now $50,000, which means your leverage has increased to 450/50 = 9, nearly double what it was before. If the value of the house declines by another 10 per cent of the original price, to $400,000, your equity goes down to zero and your leverage becomes infinite. This is a good example of nonlinearity: The second 10 per cent drop drives leverage up infinitely more than the first. We will see later in this chapter how the strongly nonlinear nature of leverage is a powerful amplifier that can cause booms and busts in the economy.

Credit causes risk. To mitigate risk, lenders usually require collateral, which in the case of a mortgage is usually the house itself. As a result, when the value of your house drops, your loan becomes riskier for the lender. When many types of financial assets, such as stocks or mortgage-backed securities, are purchased using credit, to mitigate the cost of a potential default, the lender usually requires partial repayment of the loan if the leverage exceeds a certain threshold. In the financial world this is called a *margin call*. One way to raise cash to meet a margin call is to sell some of the asset. But if many investors do this, when asset prices drop, everyone sells at the same time, lowering prices, which triggers even more margin calls and more selling, creating a feedback loop that amplifies bad news. As we'll see, sufficiently high leverage can cause bubbles and crashes to occur spontaneously for what might seem like no reason at all.

Margin calls lower the risk for an individual lender acting alone, but they increase systemic risk. In principle, it seems prudent for the lender to require a margin call; if only a few investors are using leverage, this helps protect the lender against default. But when too many investors use leverage, the collective effect of their actions can lead them all to default at the same time, causing prices to drop and thereby *increasing* the risk to the lender. Striking the right balance between individual risk and systemic risk is a key challenge for regulators. As we will discuss at the end of the chapter, some of the more progressive central banks are developing complexity-economics models to better understand how to design regulatory frameworks that take systemic risk into account.

In books and papers published in the 1970s, the Washington University Marxist economist Hyman Minsky postulated that, when markets are calm, competition drives investors to take on more leverage to increase their profits, which can cause a crash when conditions change. Minsky's arguments were not expressed in mathematical terms, and they ran counter to prevailing views about market efficiency, so they were largely ignored at the time. They proved to be prescient for the 2008 financial crisis, when Minsky's work suddenly became very relevant.

I learned a lot about leverage from my friend John Geanakoplos, whose pioneering work on leverage cycles shows how real-world experience can drive scientific discovery. John was a professor at

Yale, but he was also a financial practitioner. His cousin, Mike Vranos, was Wall Street's leading mortgage-backed securities trader at a now defunct firm called Kidder-Peabody. He arranged for John to join the firm as head of fixed-income research, where John invented new mathematical methods for pricing mortgage-backed securities. When Kidder folded (for unrelated reasons), Mike and John and four of their colleagues started Ellington Management Group, which quickly became the leading mortgage-backed-securities hedge fund.

John's model for leverage was inspired by the near meltdown of the global financial system in 1998 because of the use of excessive leverage by a high-flying hedge fund called Long-Term Capital Management (LTCM), which was started by John Meriweather. LTCM's board of directors included Myron Scholes and Robert Merton, both Nobel laureates for their work in finance. From its founding in 1994, through 1997, LTCM made remarkable returns that for two years exceeded 40 per cent. This was possible only because of their very high leverage. By 1998, they had $4.7 billion under management, but the size of their loan was an astounding $125 billion, meaning their leverage was about 25:1. To put this in perspective, with this much leverage it takes a loss of only 4 per cent (in their case $4.7 billion) to go bankrupt.

Their core strategy was a type of statistical arbitrage called *convergence trading*, which involves making bets on the spreads between interest rates on bonds, for example, by buying British bonds and selling American bonds. Their strategy was predicated on two assumptions: First, when differences in interest rates widen, they will eventually return. LTCM searched for pairs of bonds whose spreads were wider than their historical average. They would then buy the bond with the lower interest rate and short the bond with the higher interest rate (thereby betting that its price would come down), in order to make a profit when the interest rates eventually returned to normal. Their second assumption was that spreads between different bonds behave independently of each other.

Both of these assumptions unraveled in 1998. The problems

began with the Asian financial crisis in 1997. Thailand, which had a booming economy and significant foreign debt, was forced to stop pegging its currency to the dollar, after which the Thai *baht* lost half its value, and the increase in foreign debt left Thailand bankrupt. The ensuing panic affected all of Southeast Asia, causing investors to lose faith in investing in developing countries, which in turn resulted in an economic slowdown throughout the developing world. It may also have contributed to the plummeting price of oil, which declined to what was then an historical low of $11 per barrel in November 1998. In the run-up to this, Russia defaulted on its bonds. The global investment community became nervous that more such problems might follow, causing a 'flight to quality' as investors rapidly moved their capital from risky bonds to safer bonds.

This caused many of LTCM's positions to go bad at the same time. Safe bonds pay lower interest rates than risky bonds, so LTCM's convergence strategy meant that across the board it was long on low, safer interest rates and short on high, riskier interest rates. The flight to quality drove the interest rates of safe bonds even lower and the interest rates of risky bonds even higher, so LTCM lost money across the board. The fact that LTCM had become so huge and so leveraged caused both of the assumptions underlying their strategy to be violated. As problems mounted, they made the problem even worse: when spreads widened, to make their margin calls they needed to unwind their positions; but as they did so, their market impact made the spreads even wider. Because they were invested globally, this happened everywhere, causing spreads throughout the world to widen in tandem.

LTCM lost $5 billion – the fund's entire net worth – in only four months. Investors panicked: If the fund was forced to liquidate quickly, it would buy safe bonds and sell risky bonds, and spreads would widen even more.[5] Virtually every major investment bank on Wall Street had money invested with LTCM, and many had made huge convergence bets of their own, so a rapid liquidation could have pulled down the entire global financial system. The Federal

Reserve stepped in and organized a bailout, whereby LTCM's major investors were pressured into purchasing it and winding it down slowly. As in the 2007 quant meltdown, this illustrates the importance of understanding market ecology, and is a good example of how leverage leads to systemic risk.

John's firm, Ellington, also used leverage, but at reasonable levels: Where LTCM's leverage was an astronomical 25, Ellington's was only about 5. Ellington traded mortgage-backed securities, not bonds, and they were unaffected by the LTCM meltdown. However, one of Ellington's prime brokers got nervous and suddenly asked for a margin call – an immediate repayment of the loan providing the leverage – allowing only one day for repayment. Ellington saved the firm and its investors with frantic trading and has used lower leverage since then. Perhaps even more important, they changed their contracts so that their lenders need to give six months prior notice before changing Ellington's margins. This solved the real problem that had caused the margin call: Despite the fact that there were no problems with Ellington's positions in the market, increasing nervousness led the lenders to demand more collateral (forcing Ellington to sell assets).

Inspired by these events, John quickly developed a theory for how leverage causes booms and busts. He did this in 2000, eight years before the 2008 financial crisis.[6] Whereas Minsky simply presented the idea that waves of investor pessimism and optimism could influence borrowing and possibly cause a crash, John's mathematical theory linked uncertainty, leverage and asset prices, and incorporated the important role of collateral. (John was unaware of Minsky.)

To put this in context, there is a highly influential theorem by Franco Modigliani and Merton Miller which suggests that leverage shouldn't affect prices. Their theorem, formulated in 1958, is a classic result in economics, and says that the enterprise value of a firm should be unaffected by how it is financed. Enterprise value is a measure of fundamental value, not market price, and their theorem involves a long list of strong assumptions, including efficient markets, but it was widely interpreted to mean that leverage shouldn't affect prices.[7] The LTCM saga and Ellington's experience made it clear that leverage could have a *big* effect on prices.

John designed a model to show how leverage can distort prices. His model is an equilibrium model with rational, profit-maximizing agents who trade an asset (like a stock or bond). The model is formulated in a three-step process; in between each step, information is revealed that can be 'good' or 'bad'. There is a continuum of different agents, ranging from 'optimists', who have a prior belief that good information is more likely, to 'pessimists', who have a prior belief that bad information is more likely. Crucially, the agents are allowed to lend to each other.

In the first period, the model assumes markets are calm (meaning asset price volatility is low). The optimists borrow from the pessimists and buy as much of the asset as they can, pushing up its price.[8] If good news is received, this situation persists, but if bad news is received, the price drops. One expects the price to drop due to bad news, but the leverage amplifies the drop for two reasons: First, the leveraged optimists lose more money from the initial price drop than the unleveraged pessimists. This decreases the optimists' influence on the market, giving the pessimists a bigger role in setting prices, and forcing the asset price down. Second, and even more important, the uncertainty created by the news makes the lenders nervous, so they are not willing to lend as much. They lower the allowed leverage, forcing a margin call and preventing new buyers from borrowing nearly as much as the original buyers did to prop up the price. If bad news is received twice in a row, this causes the market to crash. John showed that the rise and fall in price goes hand in hand with the rise and fall in leverage, which in turn goes hand in hand with the fall and rise in uncertainty, in a process he called the *leverage cycle*. This demonstration of the dangers of leverage contradicted the prevailing wisdom at the time, but unfortunately the lesson was not absorbed in the build-up to the 2008 financial crisis.

On a broader level, John's theory of the leverage cycle provides a great example of how institutional structure can cause market inefficiency, even with rational agents. If the agents were not allowed to lend, prices would match fundamental values. The combination of heterogeneous expectations and lending drives profits up, while pushing prices away from their fundamental value. Even if the market is informationally efficient (it is not possible for anyone to make excess profits), it can be far from allocational efficiency. This is not a tiny effect – as we saw in 2008, it can be a huge effect.

Leverage causes the market to make its own news

One summer day in 2003, a young professor from Vienna named Stefan Thurner knocked on the door of my office at the Santa Fe Institute. Stefan had abandoned a career as a clarinet player for particle physics. Then, toward the end of graduate school, he decided that particle physics was stuck and got his PhD in finance instead. He subsequently started the Complexity Science Hub in Vienna, which is now probably the second-largest complex-systems research center in the world (after the Santa Fe Institute).

Stefan wanted to make a financial model and asked me if I had any ideas. I had first heard about John's theory of the leverage cycle at a talk he gave at SFI in 2001, and had been keen since then to make an agent-based model of the leverage cycle, so I suggested this as a good topic. We began a project that we would work on for a few weeks each summer when Stefan visited SFI, and during a long sailing trip we made in my boat, *Eudaemonia*, from Jacksonville, Florida, to Nassau, in the Bahamas. Since John G. was the inventor of leverage cycles and often visited SFI in the summer, as well as being a good friend, we pulled him into the collaboration.[9]

Stefan began programming a simple agent-based model. The idea was to do a complexity-economics version of John's three-period equilibrium model to see what it would tell us. John had kept his model simple, so that he could solve it with pencil and paper. We wanted, instead, to make an agent-based model to study the dynamics of leverage cycles in an open-ended and more realistic manner.

As in many earlier agent-based models (see Chapter 9), traders had a choice between a stock and a bond. But, unlike in those models, the traders used leverage, and we only had value investors; there were no trend followers. As in John's model, the value investors varied in their level of aggression (which had the same net effect as varying levels of optimism). The most aggressive funds bought the asset as soon as it was underpriced, whereas the less aggressive funds waited for more substantial mispricings. There were some other differences, but they're not important here. What distinguished our model from earlier agent-based models of financial markets was the use of leverage.[10] Our model contained a bank that could lend to the funds, but, as in real markets, there was a cap on leverage. If the leverage exceeded the cap, the funds were required to make a margin call and partially repay their loan to bring the leverage back under their limit.

We worked mainly in the summer, during Stefan's annual visits, so it took several years and quite a bit of trial and error to fully develop the model. When the 2008 financial crisis began, we realized that our model was relevant to current events, and we kicked into gear, got the code working and produced some results. We had set things up so that we could turn the leverage up or down. Without leverage, the market behaved as one would expect under rational expectations. The price fluctuated, but remained close to the fundamental value. There was no clustered volatility, and no one ever defaulted. Even in runs spanning centuries of simulated time, the largest losses were less than 10 per cent.

However, as soon as we let the funds use leverage, everything changed. As we began raising the leverage limit, clustered volatility and fat tails appeared and grew stronger. There were periods when the market was reasonably efficient; using leverage, the funds could make bigger bets, which kept the price close to fundamental values. But, from time to time, there would be a crash, triggering a volatile period in which the market was inefficient and prices strayed significantly from fundamental values. The crashes could be very large, sometimes generating losses as high as 40 per cent, and many of the funds would default.[11]

The crashes were caused by margin calls. When at least one fund exceeded its leverage limit, generating the margin call, it depressed the price of the asset even further. This raised the leverage even more, requiring another margin call and causing other funds to make margin calls, and so on. The net result was that a little bad news could be enormously amplified by the market.

We weren't too surprised by the big crashes; this is what we had been expecting. In John's model, leverage cycles were implicitly viewed as rare events that resulted in large crashes. In ours, however, there were 'crashes' of all sizes, from frequent small price adjustments to rarer, large events. Furthermore, there were long calm periods punctuated by turbulent periods – in other words, clustered volatility. This, in turn, caused fat tails in price changes, matching real markets surprisingly well.

Our model showed that there are at least two possible causes of clustered volatility in financial markets. One cause is trend following, as in the Santa Fe Artificial Stock Market or the Brock and Hommes model described in Chapter 9, and the other is the excessive use of leverage, as in our model. Usually, both are active at once, but in extreme circumstances one cause or the other can dominate. The dot-com bubble was a quintessential case of a crash caused by trend following, and the big decline in stock market prices during the crisis of 2007–2008 was a classic example of a crash caused by excessive leverage.[12]

Upon reflection, these two mechanisms are more similar than they might appear. Both trend following and margin calls cause *inverted demand functions*. As noted, normally demand goes up as the price goes down and supply goes down as the price goes down; this makes the market stable. But, when there are trend followers or when there is a margin call, *demand goes down as the price goes down*. This causes selling into falling markets, which makes the market unstable, driving prices away from fundamental values, or in other words, causing the market to malfunction.

Given our model's simplicity, its volatility dynamics were surprisingly realistic. Volatility in real markets follows a characteristic asymmetric pattern, consisting of sharp upward spikes followed by slow decay. Our model does something very similar, whereas standard mainstream models of volatility have much more symmetric profiles. Our model gave a good quantitative match to the empirical volatility profile.

One of the great values of agent-based models is that they can easily be used to test different policies. We brought in Stefan's graduate student Sebastian Poledna as a collaborator, extended the model slightly to make it more realistic, and compared three different policies for risk control. Our baseline policy assumed a constant leverage cap, as described above. We also tested two other more sophisticated, supposedly superior policies, one in which the leverage cap depends on the recent volatility of markets (when volatility goes down, the leverage cap goes up, and vice versa) and another where banks require borrowers to buy options to offset the risk of possible defaults. We tested all three policies under comparable circumstances, counting the frequency of defaults under each policy.[13]

With low leverage caps, both of the more sophisticated risk policies were successful in reducing the frequency of defaults. But with higher leverage caps, the opposite happened: The sophisticated policies made things worse, because they caused the effective leverage cap to change based on recent volatility. If volatility increased, leverage caps went down. This caused even more selling into falling markets, amplifying downward price movements, increasing volatility, lowering leverage even more, causing more selling and so forth. As a result, there were crashes that would not have occurred under the simple policy where the required leverage remains constant.

Let's reflect for a moment on how our agent-based model for leverage differs from a standard mainstream model. We determine prices in our model by matching supply and demand, so we do use the equilibrium assumption. But, in all other respects, this differs from a mainstream model. Our agents don't maximize utility. Instead, they buy or sell according to simple heuristics that approximate what fundamental investors actually do: buy underpriced assets and sell them later. This behavior is directly observable, and is unlikely to change in light of new policies. Our investors are boundedly rational, which means their views are occasionally wrong. This is why prices can deviate substantially from fundamental values and why the market crashes. It is why the market is inefficient.

Our model illustrates the paradox of systemic risk. If only one small fund uses leverage, its effect on the market is small, and margin calls protect the lender by making sure that the leverage never goes too high. But if many funds use leverage, margin calls have a systemic effect on the market, increasing the risk of default and hurting the lender as well as the borrower. Systemic risk presents a major challenge to regulators because, by its very nature, it is an emergent property, dependent on the interactions of many different actors. This means that it cannot be understood by looking at each agent one at a time – it is essential to understand the behavior of the system as a whole. Agent-based models make it easier to understand systemic risk and test policies to control it.

The financial crisis of 2008

Leverage played a central role in the financial crisis of 2008. Here are the highlights: The invention of a new type of financial instrument, mortgage-backed securities, allowed banks to bundle mortgages into standardized packages, to be bought and sold. This allowed banks to offload their risk to bigger institutions with more risk-bearing capacity. This seemed like a good thing for everyone: It was good for banks because it lowered their risk, allowing them to offer more generous mortgages, and good for low-income home buyers, because it allowed them to own homes and profit from the expected rise in prices. It also seemed good for the big financial institutions that bore the risk, because it allowed them to take on more leverage and increase profits. But in the end it was bad for everyone because mortgage-backed securities were priced on the assumption that the risk that any particular mortgage would default was independent of other default risks, whereas when things got bad, many households defaulted on their mortgages at the same time.[14]

Many consequences of this arrangement were unforeseen by most economists (who, after all, are only boundedly rational). As discussed in Chapter 4, mortgage-backed securities allowed banks to lower lending standards, which allowed more people to enter the housing market, which fueled a housing bubble. But this couldn't go on forever; after a decade of rising house prices, the pool of potential new low-income buyers was exhausted. Therefore, housing prices had to come down, and when that happened many homeowners defaulted on their mortgage payments and the value of mortgage-backed securities plummeted.

Another side-effect of offloading risk using mortgage-backed securities is that it coupled the world's financial institutions together more tightly than anyone realized. American mortgage-backed securities seemed like such a good deal that many banks and investors throughout the world bought them. The years leading up to the crisis were a period of quiet markets and steady growth. As a result, everyone was highly leveraged, so when mortgage-backed securities lost value, the portfolios holding them came under serious stress. A few financial institutions (like Lehman Brothers) collapsed or nearly collapsed and everyone quit lending. But because businesses depend on loans to keep the real economy going, this caused stock markets to drop precipitously and unemployment to skyrocket. Closing the loop, even more homeowners defaulted on their payments, and housing prices dropped further, stressing lenders and the economy even more.

To sum up, the crisis of 2008 was caused by a self-reinforcing feedback loop involving the interaction of housing markets, credit markets and the real economy. This feedback loop was not properly understood beforehand; though many pundits had predicted the housing bubble, very few of them predicted that its demise would cause a complete shutdown of credit (and hence a major recession). This was a failure to understand systemic risk.

Later in this chapter, we'll discuss a simple model that captures an important aspect of the feedback loop between volatility and credit that drove the crisis, but first we need to understand something about the modern history of risk management.

The evolving landscape of risk management

We have to control risk, but risk management can have unintended systemic side-effects, and if these aren't properly considered, the results can be disastrous. The organization that recommends standards for controlling risks in banking is the Basel Committee, which has representatives from all the G20 countries as well as important banking centers such as Singapore and Hong Kong. One of their principal recommendations concerns leverage. Starting in 1988, under what is now called Basel I, limits were mandated for bank leverage.[15] The size of the leverage caps varied based on the asset and could change from time to time.

Between 1991 and 1999, when I was at Prediction Company, conventional wisdom about risk control evolved substantially. In 1991, most firms managed risk by using fixed limits on leverage, even if they didn't explicitly follow Basel I. Even at the time, it was obvious that this method suffered from several problems.

The first problem was that the riskiness of assets can change due to clustered volatility. An asset whose price hardly moves, and so seems quite safe, might suddenly become volatile and therefore much riskier. Assigning risk based on face value does not automatically adapt to changing conditions and requires monitoring and manual adjustment.

The second problem is that price movements of assets are correlated. Positively correlated assets tend to move together, while negatively correlated assets move in opposite directions. A portfolio of positively correlated assets is riskier than a portfolio of negatively correlated assets. Such correlations mean that the risk depends on the whole portfolio and cannot be managed correctly on an asset-by-asset basis. Asset-specific fixed-leverage caps cannot take this into account.

The third problem occurs for trading strategies that hedge risks by short-selling (which produces profit if an asset drops in price). Fixed-leverage caps fail to take the resulting risk into account; again, risk needs to be managed at the level of the whole portfolio.

Because of all these problems, in the 1990s most firms abandoned fixed-leverage caps for a method called *Value-at-Risk* or *VaR*. This involves managing risk at the level of the portfolio, based on the probability of a loss of a given size. For example, on a daily scale, a 5 per cent VaR of $1 million means that 5 per cent of the time (that is, one day out of twenty) you should expect to take a loss of $1 million or more. VaR tracks volatility; if the estimated volatility increases, VaR increases, and vice versa.

In theory, VaR solves all three of the problems raised above, and it initially seemed like a big improvement over fixed-asset-level leverage caps. Volatility is far more predictable than the direction of price movements, and while volatility forecasts aren't perfect, they're pretty good under normal conditions (though of course it is the surprises that cause problems). Since VaR rises and falls with volatility, this makes risk control *dynamic* – that is, the suggested leverage varies based on market conditions. Under VaR, as volatility goes up, leverage goes down, and vice versa.

In 2004, the Basel committee implemented a new set of recommendations called Basel II, which, among other things, specified VaR as the standard for risk control.[16] This was by then accepted wisdom in the financial community. From the point of view of an individual institution, it makes good sense. When volatility is high, things are riskier than normal, so it's appropriate to use less leverage. Similarly, if volatility is low, it should be safe to use more. What could be wrong with that?

There were a few voices in the wilderness, such as Charles Goodhart and his collaborators, who worried that Basel II might have dangerous side-effects.[17] As we'll see, their concerns were valid.

Prudent risk management fueled the crisis

The 2008 financial crisis was complicated, but some of its essential elements can be captured in a simple agent-based model. This model suggests that the crisis was in part caused by misunderstanding the systemic risk inherent in Value-at-Risk. We'll see how a simple model of risk management, using the procedure recommended by Basel II, can lead to an endogenous leverage cycle, with a long, slowly building boom followed by a precipitous crash. This happens even without any shocks from the outside world, and provides a good example of how bounded rationality (based on widely used heuristics) leads to a chaotic attractor with endogenous dynamics.

The role of leverage in the crisis is illustrated in Figure 12, which shows the evolution of three important market indicators from

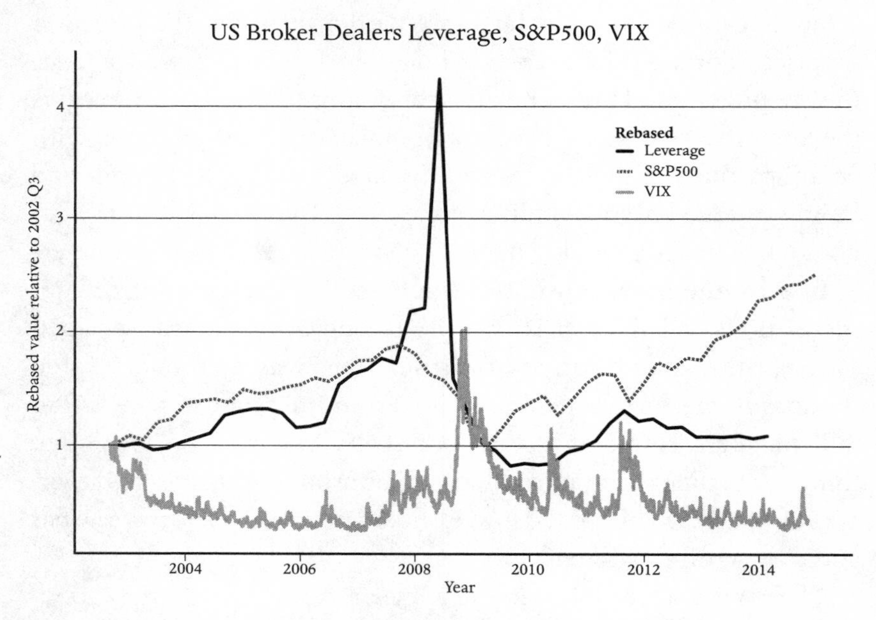

Figure 12. *Comparison of the US stock market (dashed), broker-dealer leverage (solid) and volatility (jagged).*[17]

2002 to 2015. The S&P 500, a weighted index of the largest 500 US stocks, represents the stock market. In the period leading up to the crisis, the S&P index rose steadily from a trough of 776 in October 2002 to a peak of 1,565 in October 2007, then plummeted to 677 in March 2009, a level only 43 per cent of its value at the peak and lower than its value in 2002. Such a large drop in the market capitalization of US companies reflected larger problems with the national economy, and was one of the main symptoms of the crisis. Remarkably, as the stock market was climbing before the crash, the volatility of the S&P index steadily dropped by more than a factor of 3, reaching its trough in January 2007. (This is represented by the VIX, an index of stock volatility based on options prices.) Though still low, volatility then rose slowly until it spiked in August 2008, reaching a level nearly twice as high as its starting point.

US broker-dealers exhibited the most dramatic behavior of all. A broker-dealer is an institution that trades both for its customers (acting as a broker) and for its own account (acting as a dealer). Most of the investment banks that played central roles in the crisis, such as Lehman Brothers and Bear Stearns, had broker-dealers using leverage. Broker-dealer leverage steadily climbed alongside the S&P index between 2002 and 2007, rising by roughly the same amount. When the S&P index started to fall, leverage climbed dramatically, spiking sharply upward at the beginning of 2008 and reaching a peak nearly four times its 2002 level. It then plummeted just before volatility spiked. Leverage continued to drop, so that in 2009 it was lower than it had been in 2002.

Were the movements of these indicators unconnected, or did they have causal relationships that give insight into the crisis? John G.'s model suggested they were connected; ours, however, provides an even simpler explanation that helps us understand not just the crash, but the build-up to the crash.

The model we built to answer these questions was developed by Christoph Aymanns, my first graduate student at Oxford. He had a master's degree in physics from Imperial College in London, where he was nearly top of his class, and then went to work at McKinsey

& Company for a year before coming to Oxford. The research we did was supported by a large grant from the European Union, whose projects are identified by acronyms. Ours was CRISIS, which stood for Complexity Research Initiative for Systemic InStabilities. I was scientific coordinator for this complicated grant, which involved eleven institutions spread across seven countries and was not easy to manage. One of our goals was to make a comprehensive model of a financial system so that we could test policies for averting crises.

Christoph and I set out to model how investment banks use leverage and how this affects prices. We began by building an artificial economy – an agent-based model with a financial system consisting of several investment banks and a few assets, and a simplified real economy that had production, labor and consumption. Then Christoph began playing with the model to see what it did.

Initially, the simple artificial economy we built was unregulated – that is, we didn't impose any risk-control measures on the banks, nor did they impose any on themselves. We just gave them an initial endowment of assets, which we let them buy and sell. They could use leverage in a completely unregulated way, which we called *passive leverage*. An example of the behavior of the stock market under passive leverage is shown in the top panel of Figure 13.

Over short time spans, the stock-market index varied in a random-looking way. While this resembles a real stock market, such behavior is not viable in the long run. The artificial banks in our simulation were using leverage and taking risks. In the absence of any risk control, they would eventually take risks that were too high, and they would fail.

To manage their risk, we imposed the Basel II regulations. This raised and lowered their leverage limits based on the stock-market volatility. The behavior of the economy changed dramatically, as seen in the lower panel in Figure 13, labeled *active leverage*. The stock-market index began to oscillate! During each oscillation, prices slowly rose and then abruptly crashed, before the cycle repeated itself. The oscillations varied in period and amplitude,

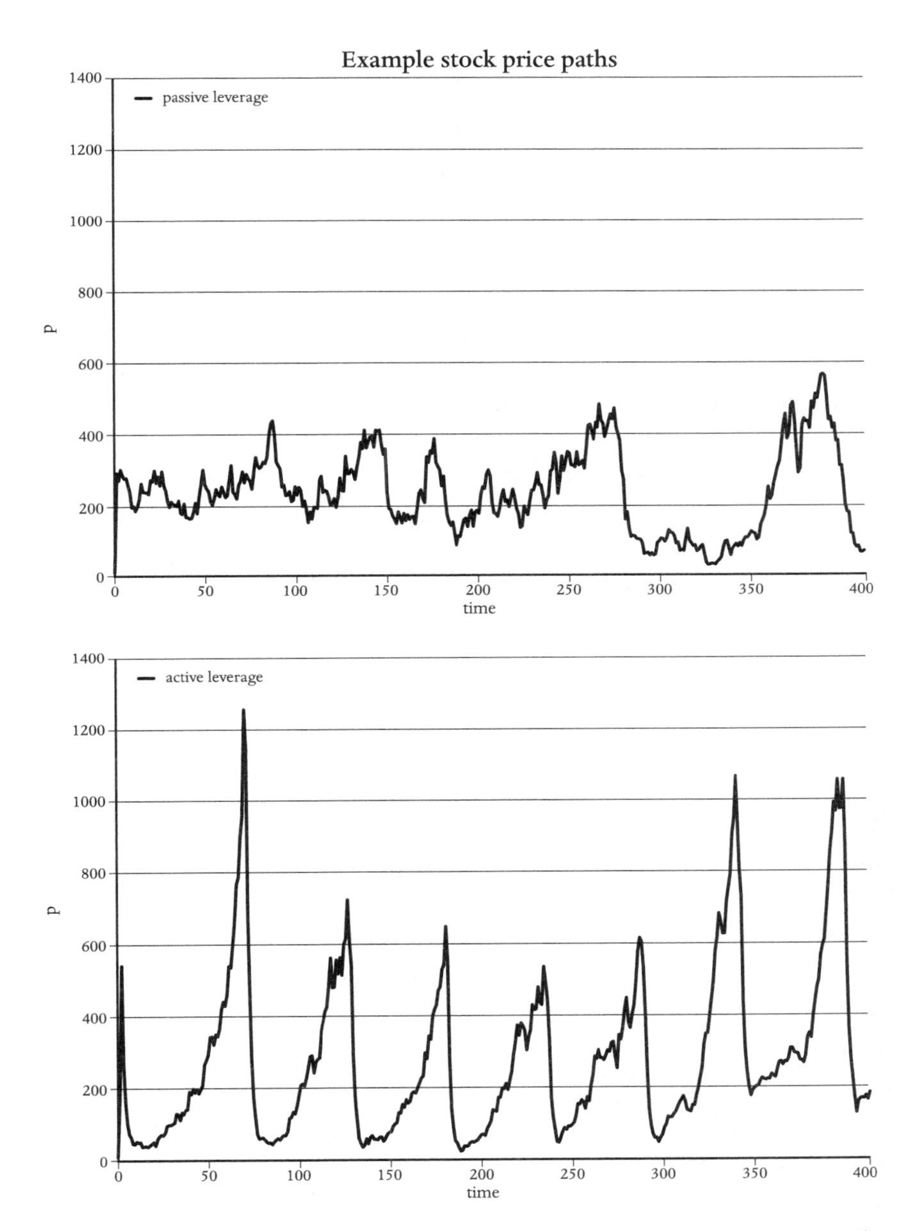

Figure 13. *Comparison of two agent-based simulations of the stock market.*[18] The panels show prices vs time for two different risk-control protocols. Given 'passive leverage' there were no risk controls. The bottom panel is exactly the same, except that the banks are required to manage their risk according to the policies recommended by Basel II.

but the basic pattern was clear. For reasons that were initially mysterious to us, the act of imposing risk control under VaR had completely altered the behavior of the market. We had imposed no shocks from outside the system; the market was making its own booms and busts, apparently for no reason at all.

To understand this behavior, Christoph began simplifying the model by removing elements until the oscillating behavior stopped. The simulation shown in Figure 13 had three banks and three stocks. Christoph reduced this to one investment bank and one stock, and even got rid of the real economy, so there was only a very simple financial system. As long as the bank managed its risk using VaR, as recommended by Basel II, we saw the oscillations.

The outcome was a model that we described in a paper called 'Dynamics of the Leverage Cycle'.[20] This model was extremely simple, which was nice for conceptual understanding but perhaps too simple to be convincing. We later teamed up with Fabio Caccioli and Vincent Tan to create a more realistic model, detailed in another paper, 'Taming the Basel Leverage Cycle'.[21] In the interest of brevity I'll skip the first model and describe the more realistic second model.

It involves two types of investor: An investment bank that uses leverage and a fund that does not. They trade a single asset, whose supply is fixed – their alternative is to hold cash. They are both given an initial endowment of the asset, and then they begin to trade. At each step, both the fund and the bank evaluate their demand for the asset, and the price is set so that supply equals demand. A diagram of the resulting model is shown in Figure 14.

The strategies of the two agents consist of a few simple but widely used heuristics. The fund is a value investor who knows the value of the asset, buys it when it's undervalued and sells it when it's overvalued (the first heuristic). The bank is an investment bank that has a leverage target and adjusts its leverage according to the recommendations of Basel II based on VaR (the second heuristic). The bank also periodically updates its volatility estimate by taking an average over the recent past. (The average

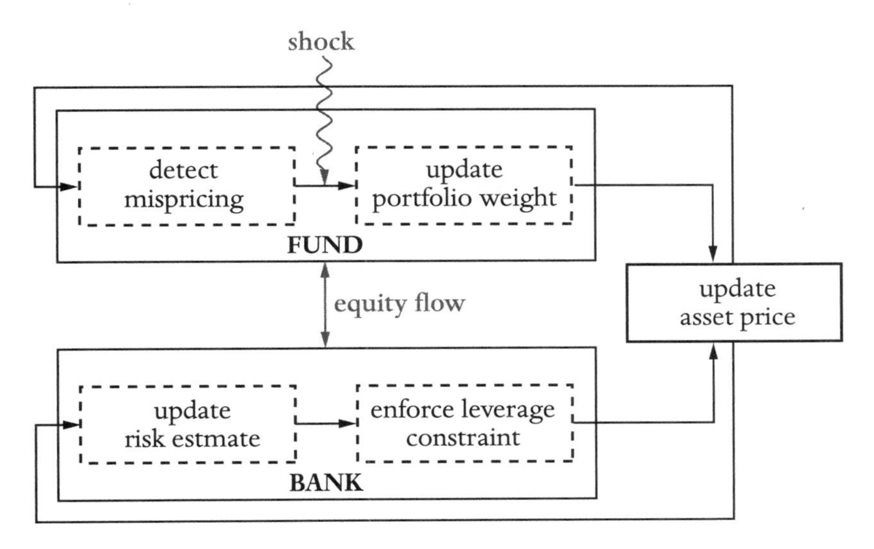

Figure 14. *A simple schematic diagram of the Basel leverage-cycle model.* There are two agents: the *fund*, which represents all unleveraged value investors, and the *bank*, which represents all investors with leverage targets. The equity flow is from their buying and selling of the asset, and the shock represents outside noise that randomly influences the fund.

is taken using exponentially decreasing weights, meaning recent volatility is weighted more than past volatility – the third heuristic.) There's also a random shock that alters the fund's estimate of the fundamental value, which can be turned up or down or even set to zero.

That's it. The model lumps the entire investment-banking world into a single bank and the entire fundamental-value world into a single fund, and ignores everything else. This is a representative-agent model, and an extreme simplification, but that was helpful here because it made the model easy to analyze and allowed us to understand why the oscillations emerged.

The model has a few parameters that control its behavior. Perhaps the most important parameter is the relative wealth of the bank and the fund.[22] If the fund has most of the wealth, then the dynamics are stable. The price of the asset stays close to its fundamental value, and the leverage and volatility are more or less constant. In contrast, if the bank has more wealth, everything changes significantly, and the system begins to oscillate, as it does in the bottom panel of Figure 13. Surprisingly, the oscillations persist even if the shocks are turned off. This is because the oscillations are chaotic; the system shows sensitive dependence on initial conditions and has all the properties of chaotic dynamics, as discussed in Chapter 1.

In the course of a cycle, the stock price, leverage and volatility behave much as they did in the crisis, as shown in Figure 12. The stock price and leverage slowly build, while the volatility slowly drops. Then the market suddenly crashes, the leverage plummets and the volatility spikes sharply upward. The frequency of the oscillations depends on the time-window used to compute volatility. When we do this using a two-year window, which is the industry standard, the cycle takes about ten years, roughly equivalent to the period of smooth growth in the build-up to the 2007–2008 crisis. The transition from the stable to the unstable regime occurs at a leverage equal to about 9.[23]

Why does the system do this? Because VaR is inherently destabilizing. Whenever the price drops, the investment bank sells in order to lower its leverage. It thus systematically sells into falling markets, and, if the bank were the only actor, the market would immediately crash. The incipient crash is prevented by the unleveraged value investor, who stabilizes the market by buying when the price falls and selling when it rises. The stability of the system depends on the competition between the value fund and the investment bank. As long as the investment bank is small or its appetite for leverage is low, the system is stable. But if the investment bank is large or its appetite for risk is high, the market begins to oscillate. This is because during the boom the estimated volatility slowly drops due to the smoothing action of averaging past volatility. As the volatility drops, the leverage increases nonlinearly, growing slowly at first and then suddenly becoming very large (recall the example in Chapter 4). Leverage acts as an amplifier; when the dial is turned up high enough, the system becomes so unstable that even an infinitesimal perturbation can trigger a huge crash. When the crash finally occurs, this drives the volatility up and the leverage down, and the cycle repeats itself.

The fact that the transition between the stable regime and the chaotic regime is both sudden and sharp poses a problem for regulators. Near the transition, a small increase in the wealth of the investment bank causes the system to abruptly move from a static fixed point to chaotic oscillations. The oscillations don't start out small and get bigger; when they appear, they are full size. This is a good example of what in physics is called a *phase transition*, in which a system suddenly changes from one state to another. More realistic versions of this model could potentially help regulators anticipate and prevent this kind of financial chaos.

When Christoph presented his model at a meeting at the Organization for Economic Cooperation and Development (OECD), it was strongly criticized by a participant from the Bank for International Settlements because of the lack of foresight of the agents. The criticism was roughly this: 'If the banks were rational, they would understand that their behavior is leading to a crash and alter it accordingly – for example, by reducing their leverage.' My response was that real people are *not* always rational, and certainly were not in the run-up to the crisis! Perhaps, if we had recently experienced a similar crisis, and if we had had good quantitative agent-based models for systemic risk in the 1990s, we might have been able to see into the future well enough to avoid such a crisis, but we didn't. Risk management was evolving during the run-up to the crisis – VaR only began to be widely implemented in the late 1990s.[24] Thus, roughly a decade elapsed from this point until the crisis occurred – consonant with the timeline our model predicts. Our model suggests that implementing a new risk-management policy *that we didn't properly understand* likely played a key role in causing the crisis. It also suggests that the use of VaR, as recommended by Basel II, together with the use of high leverage, *were sufficient to have caused the crisis on their own*. Although the model does not contain all the mechanisms that fueled the crisis, such as the housing bubble, it generates behavior that quantitatively matches the crisis and the events leading up to it. The housing bubble may only have been the spark that lit the fire.

Our model is in the spirit of Hyman Minsky's and John G.'s prescient warnings, and we all agree that high leverage is problematic. But the underlying mechanism causing problems is different: Minsky argued that greed and competition would drive leverage up when markets got too quiet, and John G. argued that the problem is due to heterogeneous investors, with different levels of optimism. We show that systemic financial risk can be caused by prudent investors – the problem exists even if investors aren't greedy, and even if they all act the same way. In our model, investors vary their leverage based on volatility because *their risk manager tells them to do so*, believing this makes them safer.

The job of a regulator like the Basel Committee is to set the rules of the game to make the financial system work for everyone. We need credit to make the economy run, but too much credit is destabilizing. Like Goldilocks, the regulator needs to get things 'just right'. That said, the precautionary principle would suggest that regulators should err on the conservative side, setting leverage limits low enough to leave an ample buffer in order to ensure stability. The regulator's job is complicated by the fact that the leverage cycle appears suddenly and at full size. Because leverage levels build slowly, and because random external events can affect prices and volatility, it's hard to know that we're in a cycle until the situation is already out of control. Nonetheless, high leverage, climbing markets and low volatility all signal that the market is in trouble.

Is it possible to formulate a better policy that allows higher leverage without increasing the risk of leverage cycles? To investigate this question, we introduced a generalization of the Basel leverage policy with a parameter that allowed us to use any mixture of Basel I (constant leverage) and Basel II (VaR). We found that the best policy depends on the situation. As we expected, when the banking sector is small and leverage is low, Basel II was the best policy. But when the banking sector is bigger and/or its leverage is higher, then the best policy is closer to Basel I. It's a matter of making the right tradeoff between individual and systemic risk.

We also explored alternative policies and found that we could significantly improve risk management by giving the bank more time to respond to changing conditions – for instance, by allowing it to adjust its leverage targets slowly when the financial system comes under stress. This caused a big improvement, and provides important advice for regulators. We could only investigate this because our model is dynamic – it cannot be understood with a static equilibrium model.

Taken together, John's original leverage model, Stefan's model presented earlier in this chapter and Christoph's model above show how standard economic theory and complexity economics are complementary. John's equilibrium model blazed the trail, showing that the Modigliani-Miller theorem does not apply, and illuminating the basic principles of leverage cycles. Stefan's agent-based model showed that leverage doesn't just lead to crashes – it contributes to clustered volatility and fat tails in price changes on an ongoing basis. Christoph's model discovered a different kind of leverage cycle, caused by risk management alone, and illustrated how this could lead to a slow boom followed by a rapid bust. Both agent-based models provided platforms to study policy, showing the advantages of alternative risk measures and the effectiveness of giving borrowers more time to reduce their leverage.

In 2010–11, the Basel Committee on Banking Supervision formulated a new protocol, Basel III, which is still not fully implemented, and there is a Basel IV on the horizon. These policies include more stringent caps on leverage and involve changes in the precise way leverage is calculated, as well as regulation of many other aspects of banking, including liquidity. When fully implemented, they will force investment banks to hold a few trillion dollars more capital in reserve than they did before, which will likely lower their profit margins. Not surprisingly, this is all very controversial. Many banks feel the regulations are excessive; other parties feel they don't go far enough to prevent the next crisis. I think the only way to resolve this question is to build carefully calibrated agent-based models that represent the heterogeneous agents of the financial system more realistically and allow us to get quantitative answers.

The construction of such models is in progress. Working with the Bank of England, my ex–graduate student Alissa Kleinnijenhuis built a prototype for a potentially realistic model of systemic risk in global financial markets, and another ex-student Garbrand Wiersema invented a mathematical method to systemically understand the channels of contagion in financial markets and their interactions, which has been used by the Bank of South Africa.[25,26] These models, which are ready to be put into practice, could provide new tools for monitoring and regulating the financial system.

In summary, providing a better understanding of systemic risk in financial markets is one of the success stories of complexity economics, which could be further developed to make financial markets safer.

The resemblance between financial and fluid turbulence

Although the financial system and fluid flow are very different phenomena, they have many things in common, and the correspondence between financial turbulence and fluid turbulence has proven useful for concrete applications such as predicting the volatility of financial markets. As we will discuss in the next part of the book, the correspondence between the two phenomena may also offer insights into how we can do a better job of monitoring and regulating financial markets.

A fluid is anything that flows freely, like a gas or a liquid. When a fluid is slightly stressed, its motion is calm (called *laminar* by the experts), but when it is strongly stressed, it becomes *turbulent*. Picture the wind blowing over the sea: On a calm day the sea is flat. As the wind starts to blow, ripples form – not only are they small, but their shape is more or less regular. As the wind blows harder, the ripples turn into waves, which get bigger and bigger, and their motion becomes more and more irregular – the motion becomes turbulent. In a hurricane this becomes what is called *fully developed turbulence*.

As Edward Lorenz realized, turbulence is due to chaotic dynamics. As soon as the movement of the ripples becomes irregular, the motion is chaotic. Turbulence corresponds to complicated chaos, with many degrees of freedom. Like financial markets, fluid turbulence exhibits clustered volatility and fat tails. In a market, the clustered volatility is due to irregular price movements; in a turbulent fluid flow, it's due to the irregular motion of the fluid. Figure 15 shows a measurement of the fluctuations of wind velocity in a given direction at a given point in space. There are periods of relative calm interspersed with gusts of wind, reminiscent of the clustered volatility for stock market returns in Figure 10.

The resemblance goes much deeper. As shown in a remarkable paper by the experimental physicist Shoaleh Ghashghaie and collaborators, the fluctuations for both fluid turbulence and finance vary across different scales in precisely the same manner.[27] They compared a fluid-turbulence experiment to a finance 'experiment'. In the fluid-turbulence experiment, they measured fluctuations in wind velocity between two probes at different locations, varying the distance between the probes from millimeters to meters. In the finance experiment, they recorded the fluctuations of the Deutschmark–dollar exchange rate by looking at the change in prices over time-intervals ranging from five minutes to two days.

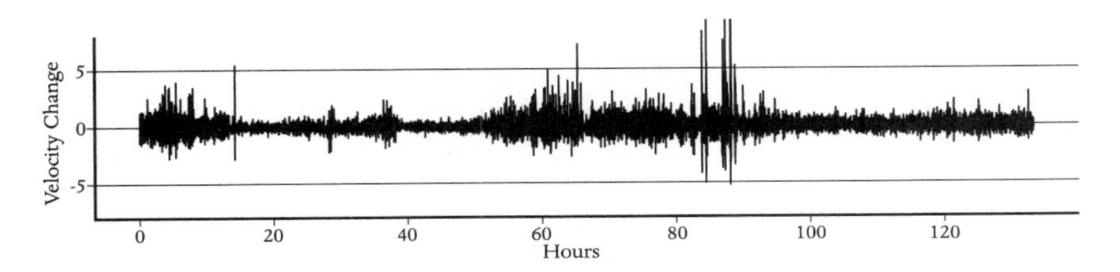

Figure 15. *Intermittency in fluid turbulence.* Fluctuations in the wind velocity as a function of time (in minutes) for turbulence in air.

Then they compared the distribution of the sizes of the fluctuations in both cases.

They first matched the millimeter-separation fluctuations for fluid velocity against the five-minute fluctuations in prices. Because of clustered volatility, both distributions have fat tails. Most of the time, the fluctuations are small, creating a sharp peak in the center, but once in a while the fluctuations become large, creating a fat tail. As they went to larger scales, either by putting the velocity probes further apart or by measuring the price changes over a longer time, the distributions became less fat-tailed. By the time the probes were a few meters apart, or the prices were two days apart, the distribution became close to a normal distribution, with thin tails. As the scale changed, the distributions of the velocity fluctuations and price fluctuations changed in exactly the same way.[28]

The analogy between fluid and financial turbulence has practical applications. This was demonstrated by a group of French theoretical physicists studying fluid mechanics, including Jean-Francois Muzy, Jean Delour and Emmanuel Bacry, who adapted their techniques for predicting volatility in fluid turbulence to finance. This required only minor changes in the theory.[29] Their models give better predictions for price volatility than previous techniques used in finance, though, as it happened, two financial economists who were then at Yale, Laurent Calvet and Adlai Fisher, independently developed a similar method around the same time.[30] (Calvet and Fisher were strongly influenced by Benoit Mandelbrot, an early pioneer in chaos and complex systems, who also knew a lot about fluid turbulence.) It is striking that the same mathematical methods provide state-of-the-art forecasts for volatility in both finance and fluid turbulence.

In Chapter 1, we distinguished between simple chaos and complicated chaos. Fluid turbulence is an example of complicated chaos. In a strongly turbulent system, like a hurricane or rapids in the Colorado River as it flows through the Grand Canyon, there is both structure and randomness. This is only a conjecture – I have no way of proving it at present – but I believe that the dynamics of global financial markets are also an example of complicated chaos, more accurately called *high-dimensional chaos* or *chaos with many degrees of freedom*.

Complicated chaos, like fluid turbulence, happens in unstable systems with many interacting components. In the financial system, when there are many agents using high leverage or when trend following becomes prevalent, prices become unstable. The global financial market has become enormously complicated and closely connected: All developed countries have stock markets and bond markets. Hundreds of thousands of assets are traded globally every day. Hundreds of thousands of investors pursue diverse trading strategies. High-frequency traders turn over their positions in fractions of a second, while low-frequency value investors hold their positions for years: The difference in timescale is a factor of 10 million![31] As we saw in the LTCM crisis and the crisis of 2008, markets do not behave independently – a disruption in Thai markets nearly brought down the US financial system. (Similarly, Lorenz famously pointed out that a butterfly flapping its wings could influence the trajectory of a hurricane.) In such a complicated system, with the strong nonlinear feedback introduced by leverage and trend following, it would be surprising if the movements of prices were *not* chaotic and high-dimensional. Financial markets are undoubtedly complicated and competitive, and as discussed in Chapter 7, chaos and clustered volatility should be expected.

Our Basel leverage-cycle model makes it clear how leverage creates nonlinearities that can cause even a very simple financial system to be chaotic. That model is so simple that it cannot exhibit complicated chaos, but in a more realistic model, replacing the two representative investors with a more realistic constellation of diverse, heterogeneous investors, one would still expect to see chaos, but with more complicated, turbulent behavior. The models in this chapter suggest that leverage in financial markets corresponds to stress in fluid flows, and indicate that as market participants use higher leverage, the financial system becomes more turbulent.

The weather is the best example of how fluid turbulence affects our lives. A concerted and coordinated global effort has allowed us to make accurate weather predictions. Could we learn from weather prediction how to better understand the economy and keep its turbulence under control? We'll explore this idea in the next part of the book.

PART IV:
Climate Economics

Climate change has already begun to alter the Earth's weather. The intensity of storms is increasing and shifts in rainfall patterns are disrupting agriculture and causing wildfires. Carbon dioxide persists in the atmosphere for thousands of years so, even if we stopped greenhouse-gas emissions immediately, we would still face a likely global sea-level rise of more than a meter this century alone, and many more in centuries to come. Moreover, runaway nonlinear feedback effects – such as the release of carbon-bearing methane from thawing tundra and less reflected light due to diminishing ice coverage – could potentially amplify human-induced climate change. We need to bring greenhouse gases under control soon to avoid catastrophic and irreversible changes in the biosphere, mass migrations, mass extinctions and disruption of the systems that provide us with food and clean water. To limit the average global temperature change to 1.5°C (2.7°F), we need to bring net emissions to zero within the next twenty years or so.[1]

While understanding climate change is the domain of physical scientists, understanding climate change's *causes and potential solutions* is largely the domain of social scientists. Greenhouse gases are a byproduct of economic activity. In order to stop emitting them, we need to transform virtually every facet of the global economy – including land-use practices, which account for about a quarter of emissions, and crucially, the way we generate and use energy, which accounts for the remaining three quarters. While the energy sector constitutes only about 4 per cent of GDP, every industry uses energy; the entire economy collapses without it.

The change from fossil fuels to non-carbon-emitting energy

sources is often called the *green-energy transition*, where the word *green* emphasizes the many other benefits, such as reduced air pollution, that will come from abandoning fossil fuels. Since renewable energy sources and fossil fuels have different supply chains, this transition will require re-wiring much of the production network, shifting the occupational balance in the labor force and building new infrastructure for generating and distributing energy. Humanity has experienced this scale of rapid change only once, during the Industrial Revolution of the late eighteenth and early nineteenth centuries. Switching to a net-zero carbon economy will also require structural transformations in other economic sectors, including transportation, construction and farming. Investors, insurance companies and central banks have begun to realize that physical risks caused by climate change, such as flooding, hurricanes, sea-level rise and lethal heat waves, pose major risks for the economy. Navigating the transition will require new business models. There will likely be significant shifts in wealth and power within and across nation-states, affecting sociological polarization and geopolitics.

Before we discuss the economics of climate change, let's look at how we learned to predict weather and climate, a story that contains valuable lessons for economics and for how we might address the economic origins of climate change.

12.

How we Learned to Predict Weather and Climate

If a single flap of a butterfly's wings can be instrumental in generating a tornado, so also can all the previous and subsequent flaps of its wings, as can the flaps of the wings of millions of other butterflies, not to mention the activities of innumerable more powerful creatures, including our own species.

Edward N. Lorenz (1972)[1]

Predicting the weather

Attempts to predict the weather date back thousands of years, but weather forecasting achieved a huge breakthrough in the second half of the twentieth century, thanks to a combination of better science, better measurements and better computers. Over the last seventy years, a coordinated, large-scale, global effort and an investment of hundreds of billions of dollars have enabled us to predict the weather remarkably well.[2] Because we make daily global weather forecasts, we test weather-forecasting models all the time, so we know how good they are. Testing climate models is more challenging, because it requires running scenarios we have not yet experienced. Nonetheless, reliable weather models give us a good starting point for modeling climate.

Understanding the weather has been a matter of concern throughout recorded human history. Babylonian tablets discussing weather forecasting date from 650 BCE. The word *forecast* was coined by Robert FitzRoy, who captained the HMS *Beagle* when Charles Darwin sailed with her. In 1854, he founded what would later be called the British Met (Meteorological) Office with the help of Francis Beaufort (whose surname all sailors will recognize). In 1870, the US Weather Bureau (now called the National Weather Service) was formed; it collected data from more than 500 observation stations and made daily forecasts throughout the country.[3]

In those days, forecasts were subjective, based on experience and intuition. Forecasters kept a large library of past local weather maps, which they studied carefully and used to develop rules of thumb (heuristics). Much as we did at Prediction Company, they observed patterns in past behavior that allowed them to anticipate future behavior. One of the procedures they followed, called the *method of analogs*, illustrates the underlying principle that these methods are implicitly based on. It works like this: The forecaster studies the current weather map, and might think, 'This reminds me of the build-up to that big storm back in December of 1935.' Then he looks through his library to find similar weather maps. If December 10, 1935 provided a good match, he would look ahead to December 11 as a possible forecast for tomorrow's weather, adjusted to account for differences in factors like wind speed and direction. He might search for other, similar weather maps and repeat the procedure. Then he would lick his finger and put it in the air, make a subjective judgment, and produce a forecast.

The basic principle of the method of analogs is that the future will follow the same patterns that existed in the past. This principle is extremely useful, and is the basis of all statistical models, with broad applications beyond weather forecasting. It underlies the breakthroughs in machine learning that are revolutionizing everything from image recognition to writing high-school term papers. In fact, such methods are now used by Google to make forecasts a few hours ahead in places where the coverage of weather models is poor.

The method of analogs, and its extensions via econometric models, is still the most useful forecasting tool in economics. Instead of weather maps, the forecaster compares economic conditions in the present to past events. This method is limited, however, because there are just too many possible patterns. There's an enormous amount of information in a weather map (or in the state of an economy) at any given point in time. Even with a library of thousands of historical maps, the weather on any given day is never really close enough to the weather on any previous day to produce an accurate forecast. (Similarly, the economic conditions now are rarely similar enough to conditions in the past to make reliable predictions.) Because the weather is chaotic, small differences between today's weather map and a historical weather map turn into large differences between them on the following day. Subjective weather forecasts were better than nothing, but their value was limited. A new approach was needed.

In 1901, Cleveland Abbe, the Weather Bureau's chief meteorologist, proposed a radically different method that has come to be called *numerical weather forecasting*.[4] He suggested using the laws of physics to predict the weather. His proposal had the same essential elements as our method for beating roulette discussed in Chapter 1: understand the forces, measure initial conditions and use them to make predictions. Of course, weather is vastly harder to predict than roulette. It involves the interaction of temperature, pressure, wind velocity and humidity, as well as many other variables, such as topography. Gathering initial conditions requires simultaneous measurements over a fine grid covering the entire globe. It now seems obvious that we should predict weather this way, but in 1901 there were no computers, measurements were limited and the physics of the weather was still poorly understood. Cleveland Abbe's proposal was regarded as audacious and utopian and led to no immediate action.

The first numerical weather forecast was published by the English polymath Lewis Fry Richardson in 1922.[5] Richardson was an early complex-systems scientist who made contributions to mathematics, physics, meteorology and psychology. He was one of the first to articulate the idea of fractals, and he was the first to develop a quantitative theory for the causes of wars. His Quaker upbringing made him an ardent pacifist, which exempted him from military service, but in England at the time, it also disqualified him from holding a proper academic appointment. Richardson's attempts at numerical weather prediction continued even while he was working in a Quaker ambulance unit during World War I. He eventually set himself a very specific task: To predict the weather for 1.00 p.m. on May 20, 1910, based on measurements made at 7.00 a.m. that day. The weather depends on nonlinear equations whose dynamics are chaotic, which means that the equations have to be solved step-by-step through time and require an enormous number of arithmetic calculations. To cope with this challenge, he introduced a procedure for solving the equations numerically by hand. I mean literally by hand: He performed tens of thousands of arithmetic calculations with pencil and paper! Even using the biggest supercomputers in the world, the time and memory such calculations take continues to define the limit on how accurately we can predict the weather. Imagine what it was like for Richardson using nothing but a pencil and paper – and on a battlefield to boot.

Making a six-hour-ahead forecast took Richardson about six months. Unfortunately, the way he ingested the data into his model was flawed, so the answers he obtained were off by a factor of 100. The publication of his failed results had the opposite effect of the one he had hoped for, putting a damper on numerical weather prediction for the next two decades.

In his 1922 book, Richardson went beyond his weather-forecasting results and wrote about how parallel-processing might be used to make his methods practical. He imagined a city of 10,000 human 'computers', who would live in a special city. Every day, they would go to work and take their positions in a stadium-like enclosure. Each human computer would perform designated calculations with pencil and paper, and pass the results to her neighbors on every side (while her neighbors passed their calculations to her). Then the process would be repeated, with each 'computer' making a new calculation in response to the inputs from her neighbors. Richardson estimated that even though each calculation would still be slow, the parallel-processing power of 10,000 people working simultaneously would speed things up enough to produce useful real-time forecasts. It was an extraordinary vision, but it was never implemented for weather prediction. It was, however, put into action during World War II, when squads of human 'computers' solved equations relevant for the design of the atom bomb and performed other calculations for the war effort.

In 1950, John von Neumann was searching for non-military applications for digital computers and decided that weather prediction was a good candidate. He spoke to Carl-Gustaf Rossby, one of the leading meteorologists of the day, who in turn recruited his young associate at the University of Chicago, Jule Charney, to lead a team of other researchers. Working intensely for two years, they produced a weather model simple enough to run on the ENIAC computer.[6] The ENIAC was extremely crude by the standards of modern computers. It had only 10 words of read-write memory and 624 words of read-only memory, and could do 357 multiplications per second – more than a million times slower than a modern smartphone. Von Neumann's wife, Klara, taught the team how to program and checked their code. By working around the clock with the ENIAC for thirty-three days, they produced four retrospective forecasts. The results were not as good as existing subjective forecasts, but they were good enough to provide a proof of principle – and the field of numerical weather prediction was born. At the time, the next day's forecast still took twenty-four hours to perform on the ENIAC, but the meteorologists were confident that computers would speed up and reduce calculation times.

The global scientific community, governments and the military saw the potential of numerical weather forecasting, and work began in earnest to make forecasts faster and more reliable. Since the 1950s, hundreds of billions of dollars of public investment have gone into producing daily weather forecasts using numerical weather prediction. This effort has included developing a better understanding of the physics involved, improving numerical methods for solving the underlying equations on a computer, getting more accurate measurements (particularly via satellites) and taking advantage of bigger and faster computers.

With hindsight, it might seem obvious that numerical weather prediction would be better than subjective weather forecasting, but it took almost thirty years of hard work for this goal to be achieved. I became aware of the contest between subjective and numerical forecasters at its turning point, around 1980, while I was a graduate student doing research on chaotic dynamics. I had the good fortune to be asked to intern at the National Center for Atmospheric Research (NCAR) with Edward Lorenz, the pioneer of chaos theory and one of the leading meteorologists of his day.

Since my internship was unpaid, to save money I camped illegally on the spectacular mesa that NCAR sits on, near Boulder, Colorado. My tent had to be hidden during the day, and because I was using NCAR's supercomputer until late in the evening, it was a nuisance to set my tent up in the dark. If I was confident it wasn't going to rain, I preferred to sleep under the stars. I told Ed about my dilemma, and in my conversation with him at the end of each day, I would ask him whether he thought it was going to rain. He had a corner office with a big window looking out over the Great Plains. He would peer at the sky and study the clouds for a few moments, and then give me one of three answers: 'Yep', 'Nope', or 'Can't tell'. He was an excellent subjective weather forecaster: With the caveat that he used 'Can't tell' about 50 per cent of the time, he was never wrong, and I never got wet due to a bad forecast. I later learned that he had been a weather forecaster in the Pacific during World War II, so he had had some practice.

I remember asking him when he thought he and the other physical meteorologists would beat the subjective weather forecasters. He told me they were breaking even, but steadily getting better and would soon pass them. This is exactly what happened – numerical weather prediction now produces far more accurate forecasts than subjective weather forecasters ever did.

The enormous improvements in accuracy have changed our whole attitude toward weather forecasting. Although reliable long-term forecasts are inherently impossible because of the underlying chaotic nature of the physics of fluid turbulence, short-term weather forecasts are now reasonably accurate. This is economically useful: Estimates indicate that the US spends about $5.1 billion per year on weather forecasting, for an economic benefit of roughly $30 billion.[7] Since 1980, forecasting accuracy has improved by about one day per decade – that is, the six-day forecast now is about as accurate as the five-day forecast was ten years ago.[8]

The standard economic models that many national statistics offices, central banks and treasury departments use to make economic predictions are on a much smaller scale than weather forecasts, and they aren't nearly as useful. We're still at the point where the method of analogs and its econometric extensions perform as well as or better than theory-based models. As discussed, standard models operate at a much higher level of aggregation and do not make use of microdata. If weather models have an economic value to the US of $30 billion per year, better economic models could save us trillions of dollars a year, but there is no large-scale effort to build such models.

Predicting the climate

The global-circulation models that are used to predict the response of the Earth's climate system to increased levels of greenhouse gases were developed out of models for weather prediction. Given that the dynamics of the weather are chaotic, and therefore unpredictable in the long run, you might think that modeling the climate, which involves predicting on timescales of centuries or more, would be impossible. Fortunately, climate involves a different kind of prediction, which makes it in a sense easier: Climate models make a prediction about the statistical properties of the weather rather than a forecast of its future state. We can't say whether or not it will rain a hundred days from now; we can, however, estimate how often it will rain in a world with more greenhouse gases by running a numerical weather simulation with increased carbon dioxide in the atmosphere for a very long span of time and recording how often it rains in the simulation. Similarly, we can predict how high, on average, global temperatures will rise over time in response to a given level of carbon in the atmosphere. Thus, while chaos imposes a fundamental limit on short-term weather forecasting, it imposes no such limit on the accuracy of climate prediction.

One of the best ways to test climate models is to use them to predict the past. Carbon-dioxide levels and temperature have both varied substantially over time. Tests are made by running global-circulation models with past carbon concentrations to see whether the resulting temperature predictions match the corresponding historical values. The match is still far from perfect, but substantial progress has been made.

Climate models have improved substantially in the last forty years, but their predictions are surprisingly consistent with earlier models. In 1979, Jule Charney led an early study of climate change entitled *Carbon Dioxide and Climate: A Scientific Assessment*.[9] The report stated, 'We estimate the most probable global warming for a doubling of CO_2 to be near 3°C with a probable error of ± 1.5°C.' Almost precisely the same statement appeared in a report of the United Nations Intergovernmental Panel on Climate Change thirty-five years later.[10] Our current carbon-dioxide level is about 1.4 times the pre-industrial level, and unless we take drastic action immediately we are likely to reach the doubling point in about fifty years. The last time the Earth was 3°C degrees warmer than it is now was 3 million years ago, when beech trees grew in Antarctica, the seas were 80 feet higher, and horses galloped across the Canadian coast of the Arctic Ocean.

It is a remarkable tribute to the power of science and the usefulness of large-scale models that we can forecast the complex phenomenon of climate change fifty to a hundred years into the future. But climate models can inform us only about the consequences of our actions; they don't tell us how to solve the problems we've created. Re-wiring the economy so that it no longer emits greenhouse gases will require a major structural transformation. Our current economic models are woefully inadequate to guide us through this transition. If only we had economic models as good as our climate models . . .

Lessons for economics

There are several striking differences between the way we make economic predictions and our predictions of weather and climate:

> *Weather forecasts have systematically improved over time.* Economic forecasting, in contrast, has never been very accurate and has made only slow, incremental progress over the last eighty years. Numerical forecasting of the economy has yet to convincingly surpass simple statistical models implicitly based on the method of analogs.
>
> *Weather is modeled from the bottom up.* Weather modelers seek to model the atmosphere at the finest scale possible, whereas government models for policymaking operate at the aggregate level of a nation-state and use only a single representative household. The contrast with weather models is stark: It would be utterly hopeless to predict the US weather based solely on measurements of its average temperature, barometric pressure and wind velocity.
>
> The fact that our ability to make measurements and run models at a finer scale has been the biggest driver of improvements in weather forecasting's predictive power demonstrates the importance of a bottom-up approach. Current weather models track the relevant variables on a three-dimensional grid in which each cell is 10 kilometers by 10 kilometers wide and 100 meters high, with roughly 100 such cells stacked vertically. Each cell requires a measurement of temperature, wind and humidity, which

makes for a total of about 10 billion measurements over the entire globe. The global-circulation models for climate use a coarser grid, 100 kilometers by 100 kilometers wide and 1 kilometer high, but this still produces an enormous amount of information. Just as the improvements in computers over the last seventy years have been driven by the increasing density of integrated circuits, many if not most of the advances in meteorology have been driven by the increased density of grid points in weather measurement and simulation. Weather simulation using the finest grids possible requires state-of-the-art computers, and weather prediction employs some of the largest computers in the world, each of which consumes about 10 megawatts of power (enough to supply about 8,000 American households).

Weather models are highly nonlinear. Weather is an emergent phenomenon, and a linear weather model would be useless. In contrast, while economic dynamics are also an emergent phenomenon, and most of the DSGE models we talked about in Chapter 5 are intrinsically nonlinear, many of those used for policymaking are simplified and replaced by linear approximations that are easier to solve. This means, among other things, that endogenous business cycles are ruled out from the get-go.

Weather models are global. Weather models are *closed* – that is, they cover the entire world, so that a storm never appears from 'off the map'. This is essential to capture endogenous phenomena. In contrast, most official economic models are *open* – meaning, for example, that they pertain only to a single nation-state, so that outside influences from other countries (which are clearly important, such as a default in Thailand) have to be treated as exogenous shocks. While some types of economic models are global, many, if not most, still model only one country at a time.

If we were to implement agent-based economic models at a global scale, similar to those used for weather or epidemiology, all of the 'in contrast' statements above would disappear. Such agent-based models would use the finest scale feasible. With an all-out effort, it is now conceivable to construct agent-based models on a one-to-one scale, so that each simulated person in the model corresponds to a similar real person and each company corresponds to a specific real company.[11] Running such a model, which would require state-of-the-art data gathering and computer power, would cope with nonlinearities and thus could potentially predict such emergent phenomena as recessions, financial crises, inequality and economic growth. These predictions could be made at many levels, from individual firms and groups of households to industries, nations and the whole planet. While there are factors that make simulating the economy harder than simulating the weather, there are also factors that make it easier.[12]

Complexity economics can do for economics what Cleveland Abbe imagined for weather prediction. What Abbe proposed took eighty years to reach fruition, but today we complexity economists already have adequate computer power and much of the data needed to realize our ambitions. With an appropriate effort, we could build much more accurate agent-based models in five years or less, which could give us invaluable guidance in negotiating the climate-driven transition of the economy and other global problems that depend on economics. In Chapter 14 I will briefly describe some efforts that are already underway.

13.

Climate Economics and Technological Progress

Sensible policies on climate change should weigh the costs of slowing climate change against the benefits of slower climate change.

William Nordhaus (2003)[1]

Predictions about the anticipated impacts of climate change aren't perfect: We can't say precisely how much warming will happen for a given level of carbon in the atmosphere, and we don't know exactly how high and how quickly sea levels will rise, or when the carbon locked in the tundra will warm enough to begin emitting vast quantities of methane. However, there is a remarkable scientific consensus that these effects present a grave danger. The economics of climate change, meanwhile, remains highly controversial – both as to how much the green-energy transition will cost and the right way to make it happen. Some say that it will be very expensive, while our recent work (discussed below) says it will save us money. Some say we need a crash program to build more nuclear reactors; others (like me) say this would be a waste of money that could be spent more usefully elsewhere. Some say we need a carbon tax or a cap-and-trade system; others say we should simply criminalize carbon emission. Some say that, with the right incentives, the market will take care of everything; others say the government should play a leading role – for example, by investing in new technologies. Some say that we must abandon the idea of economic growth; others (like me) say that we can stop emitting greenhouse gases while continuing to grow if we invest in the right technologies. Far too many people say there is nothing we can do. To make the green-energy transition rapidly and cheaply, we need reliable models, based on empirical data, that allow us to predict the likely outcomes of each possible course of action.

To reduce and ultimately eliminate greenhouse-gas emissions produced by the burning of fossil fuels and bad agricultural practices, we need to dramatically change the technological underpinnings of much of the global economy. What is the best way to do this? How much will it cost?

I realized that I could contribute to solving this problem when the Director of the US Department of Energy's National Renewable Energy Laboratory, Dan Arvizu, came to the Santa Fe Institute in 2008 with a few of his scientists and asked us to help them 'think out of the box'. The core of the planning problem was to understand which green technologies were most likely to be able to replace fossil fuels soonest and at the most affordable price.

Our conversations led to a workshop at SFI attended by leading technologists and economists studying the climate transition. The room was full of disagreement. Many advocated for wind and solar energy, together with various forms of energy storage. Others were fans of nuclear power; still others advocated carbon capture and storage – that is, removing the carbon dioxide when fossil fuels are burned, before it enters the atmosphere. Who was right? It was not enough to know what different methods of generating energy currently cost; we needed to know what the costs would be twenty to fifty years in the future. The economics of climate change hinges on anticipating how the costs of the various technological solutions are likely to change over time.

William Nordhaus, an economist from Yale, was present at the workshop. In a paper published in 1991, he was the first person to address the economics of climate change.[2] In February 1992, he unveiled the Nordhaus model, DICE, which stands for Dynamic Integrated Climate Economy. It is based on a traditional, aggregate macroeconomic model, similar to the Solow model discussed in Chapter 5, using a single representative household that works for a single representative firm that produces a single good. Nordhaus assumes this represents the entire world. His DICE model extends the Solow model by adding an *abatement function* and a *damage function*. The abatement function specifies the cost of reducing emissions to a given level. You can think of abatement as an investment – by spending a given amount of money every year, we reduce emissions by a certain amount. The damage function gives the future economic cost of global warming as a function of total emissions. The reduction in the cost of the damages is like the return on the investment. Nordhaus performs a cost-benefit analysis: The right level of investment to combat climate change is computed by maximizing the utility of the representative agent, which is effectively the best compromise between the cost of abatement and the benefit of avoiding future damage.

While the Nordhaus model recommended that we should take action to mitigate climate change, it suggested that we should do so at a measured pace. In a recent paper, Nordhaus wrote that the optimal, 'welfare maximizing' course of action would eliminate emissions so slowly that it would result in about 3.4°C (6.1°F) of warming above pre-industrial levels.[3] That's a lot of warming! Though the resulting sea-level rise would occur slowly, it would eventually submerge most large cities. Nordhaus subsequently received the Nobel Prize in 2018 for this work.

I strongly disagree with Nordhaus's conclusions. Many prominent mainstream economists, such as Nicholas Stern, Joseph Stiglitz and Jeffrey Sachs, also disagree.[4] They argue that Nordhaus's abatement function is much too costly, his damage function is much too low, and his discount rate is too high (meaning he doesn't give enough weight to the future in relation to the present). I agree with these critiques, but I go a step further: *Addressing energy-related emissions to combat climate change will very likely save us a lot of money*. The technological forecasts from my research group at Oxford indicate that the green-energy transition will ultimately make energy prices substantially lower than they have ever been. From a purely economic perspective, even if climate change were not a looming threat, we should still make the energy transition quickly (doing most of it within twenty years). In addition to the huge benefits a fast transition will reap for the climate, if we discount all the costs to the present, it will probably save us tens of trillions of dollars. Here's how my collaborators and I arrived at that conclusion.

Predicting technological progress

For the economy to make the climate transition as quickly and cheaply as possible, we need to know which technologies are most likely to succeed in replacing fossil fuels quickly at the lowest cost. Given that technological progress depends on innovation, which is by definition unanticipated, predicting the pace of progress might seem a senseless task. But surprisingly, rates of technological progress are very predictable, even if the specific technological solutions that will achieve these rates are not.

Forecasting technological progress is possible because technological change is persistent, and the rates of change for different technologies vary widely. Figure 16 documents the enormous diversity in prices of a few representative goods and services in the United

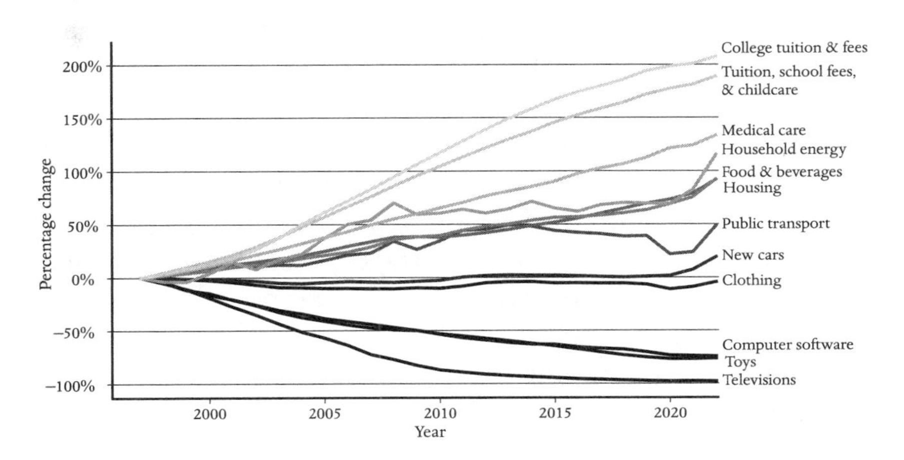

Figure 16. *Differences in the price of some goods in the US from 1998–2018.*

States from 1998 to 2018, with prices adjusted to include inflation and taking quality improvements into account. Lower prices are one of the most salient manifestations of technological progress. In those twenty years, the price of hospital services skyrocketed, increasing by more than a factor of 2, while the cost of televisions dropped by a factor of 20. Hospitals made negative progress, at least in terms of the cost to society, while televisions made rapid progress.[5] Furthermore, the rates of change are persistent: If we had observed these trends for the first ten years, we could have made a good prediction for the next ten.

Moore's Law is perhaps the best-known method of predicting technological improvement. In 1965, Gordon Moore, Director of R&D at Fairchild Semiconductor, noted that the density of components in integrated circuits doubled roughly once a year and predicted that this trend would continue for at least ten more years.[6] In 1975, he revised his prediction to a doubling in density every two years. Roughly fifty years later, his prediction has proved remarkably accurate. While his prediction applied only to the density of components (transistors, capacitors, resistors, etc.), similar improvement rates hold for computer cost, speed and energy consumption per bit of computation.

It turns out that many other technologies follow a generalized version of Moore's Law, in the sense that their prices fall roughly exponentially with time. This was demonstrated in a 2006 paper by Heebyung Koh and Christopher Magee. Chris Magee attended the 2008 SFI workshop on green technologies. He is an emeritus professor of engineering systems at MIT and was previously Director of Programs and Advanced Engineering at the Ford Motor Company. Koh and Magee gathered data on a variety of different technologies, such as electrical transmission cables, flywheels and batteries, and demonstrated that these technologies improved at exponential rates over time spans of many decades, though with highly diverse rates of progress.[7]

Gordon Moore was not the first to propose a law for technological change. There is a much older law for technological improvement that goes under a variety of names, including *learning by doing*, the *experience curve* and Wright's law. This law was originally postulated in 1936 by Theodore Wright, who has an interesting background. He was the brother of both the influential evolutionary biologist Sewall Wright and Quincy Wright, one of the pioneers of international political science. Rather than following his brothers into academia, he became a World War I flying ace and went into the aviation business after the war. In 1936, he observed that whenever the cumulative production of a given type of airplane from a given factory doubled, its cost of production dropped by about 20 per cent.[8] As it turns out, this law applies to many other technologies besides airplanes, though the rate at which the price drops varies substantially from technology to technology (and in most cases is close to zero).

Wright's Law applies not just to products emerging from a specific factory; in many cases it applies globally as well, meaning that when the global production of a given technology doubles, its average global cost of production drops by x per cent, where x depends on the technology. Wright was a special advisor to Roosevelt during World War II, in charge of boosting aircraft production, and he successfully used his own law to forecast the costs of airplane manufacturing. However, for most technologies, x is close to zero, which means that Wright's Law is not useful. For some technologies, like nuclear power, x can even be negative (meaning the cost goes up over time rather than down).[9] Because of the wide variation in improvement rates, supporting some technologies is far more effective than supporting others.

Why should Wright's Law hold at all? The popular hypothesis is that cumulative production correlates with level of effort, and improvement is proportional to effort; as more of something is produced, more effort will be put into improving it, and costs will drop. In a paper led by my ex–graduate student and now Santa Fe Institute colleague James McNerney, we explored the hypothesis that engineers improve things using a zero-intelligence model, in which they try out random solutions and pick the best ones. We showed that this leads to Wright's Law, and that complicated technologies with many interacting parts should improve more slowly.[10]

Wright's Law and the generalized Moore's Law are important because they provide the best tools we have for predicting the future performance of a technology. In a commentary on nuclear power in *Nature*, in 2010, I published my first prediction of solar-photovoltaic costs using Moore's Law, predicting that it would be cheaper than coal-fired electricity by 2020.[11] To put this in context, at the time the global weighted average cost of solar-PV energy without storage was about 38 cents per kilowatt-hour (in 2020 dollars), much higher than it is now, and most commentators regarded my prediction as implausible. A widely quoted 2014 article in *The Economist* held that 'solar power is by far the most expensive way of reducing carbon emissions'.[12] Since then, its cost has plummeted, even more steeply than I expected; I predicted 12 cents per kilowatt-hour by 2020, and instead it was only 6 cents.[13] Now solar energy is substantially cheaper than nuclear power, whose costs are in the 12–20 cents per kilowatt-hour range. In most locations, solar energy is now even cheaper than coal-fired electricity.

Okay, I was more-or-less right, but so what? My arguments were not *convincing*. To be convincing requires more evidence. We needed to gather data and perform statistical tests.

Our early data-gathering efforts at SFI were led by my Transylvanian postdoc Béla Nagy. Béla was then in his forties; he had had a career as a computer programmer before getting his PhD in statistics from the University of British Columbia. He was passionate in his quest to understand technological progress, and he worked hard to make our project succeed.

Given that technological progress is the most important driver of economic growth, you might think there would be extensive databases documenting it. Quite the opposite: At the time, it was difficult to find data systematically recording the performance and cost of technologies over long spans of time. Government statistical agencies collect data on consumer products and intermediate economic goods, but only for economically significant products, and such data is often too aggregated to be useful. We wanted data for key technologies and tasks, such as transistors or gene-sequencing – data that statistical offices do not track. So we had to collect our own, but we had no budget to do so. Béla solved this problem by assembling a group of high school and college students who devoted free time to help him scour the literature. They even wrote to Gordon Moore, who was kind enough to send us the data he had personally collected on electronic components. The result was the Performance Curve Database, which has data for about 100 technologies, covering a variety of performance metrics, including cost, as a function of time.[14] Though created more than a decade ago, it is still the only publicly available database documenting technological progress.

To test the predictive power of Moore's Law, Wright's Law and several other proposed laws for technological improvement, our research team pretended we were in the past. We then made forecasts about future improvements using only the information that would have been available back then and compared our forecasts to what really happened. We used this method to measure the forecasting accuracy of six different hypothesized laws for technological progress.[15] We made about 6,000 forecasts for sixty-two different technologies over time horizons ranging from one to thirty years into the future. To our surprise, Moore's Law and Wright's Law tied for first place. Their forecasts were better than any of the other four alternatives, and they finished the race in a dead heat.

When we looked more closely at the data, we realized that there was a simple explanation for the tie: For almost all of our technologies, the cost dropped roughly exponentially as the cumulative production increased exponentially. A simple math exercise shows that this results in Wright's Law. Béla later realized that this equivalence had been pointed out by Devendra Sahal, an Indian scholar of technological change who wrote four books about technological progress in the 1970s and early 1980s.[16] But when Béla tried to track him down, he found that Sahal had mysteriously disappeared after the last of these publications. Béla became obsessed with Sahal, gathering every possible piece of information about his life and corresponding with his brother in India. Then, about a year later, Béla also mysteriously disappeared. I hope that he and Sahal found each other, and that they are together somewhere, discussing the causes of technological progress.

One of the goals of my work with Béla was to find a formula for the accuracy of Wright's Law and Moore's Law. Making forecasts of unknown accuracy, which was the usual practice prior to our work, is of limited usefulness – if you don't know how good a forecast is, how can you rely on it? Béla found an approximate formula, but it was complicated and didn't work particularly well, and we had no good way to test it.

After moving to Oxford, in 2015 I hired a postdoc named François Lafond, and we found an elegant solution to this problem. François is an economist by training, an extraordinarily open-minded scholar, and among the most intellectually honest scientists I know. He is skeptical of everything, particularly his own work – which means his results can be trusted. I have so far managed to convince François to stay at Oxford, and he now co-leads my research group.

François and I derived a simple formula for the accuracy of Moore's Law. We began by generalizing the law, reformulating it as what is called a *geometric random walk with drift*.[17] Our underlying hypothesis was so simple that we assumed that our formula would give mixed results, but when François tested it on the Performance Curve Database, to our astonishment, it worked extremely well. When updated to allow for year-to-year correlations, it did a good job of predicting the accuracy of our forecasts for diverse technologies and time horizons. With a little help from me, François managed to derive a similar formula for Wright's Law.[18]

Our formulas for Moore's Law and Wright's Law go beyond the state of the art by predicting the *distribution* of possible future outcomes. Unlike the simple application of Wright's Law, which gives a point forecast with a single answer of unknown reliability, our formula gives the likelihood of each possible outcome using a method that has been extensively tested on many technologies. For example, our forecast for solar panels using Moore's Law is shown in Figure 17. The most likely cost for 2030 is substantially cheaper than it is now, but there is still a 5 per cent probability that solar panels will be at least as expensive as they are now. Lower costs are likely but not guaranteed. This forecast was made in 2013; costs since then have dropped even faster than the median prediction, but they're well within the likely values we predicted. Our confidence in the formula comes from the fact that we know it has done well in the past, not just for solar-PV modules but also for other technologies.

Our method allows technologies to differ in their rates of improvement. It also allows them to differ in their volatility: A technology that has historically stayed close to its trend line is more predictable than a technology that strays from it (as solar modules did during the first decade of the millennium).

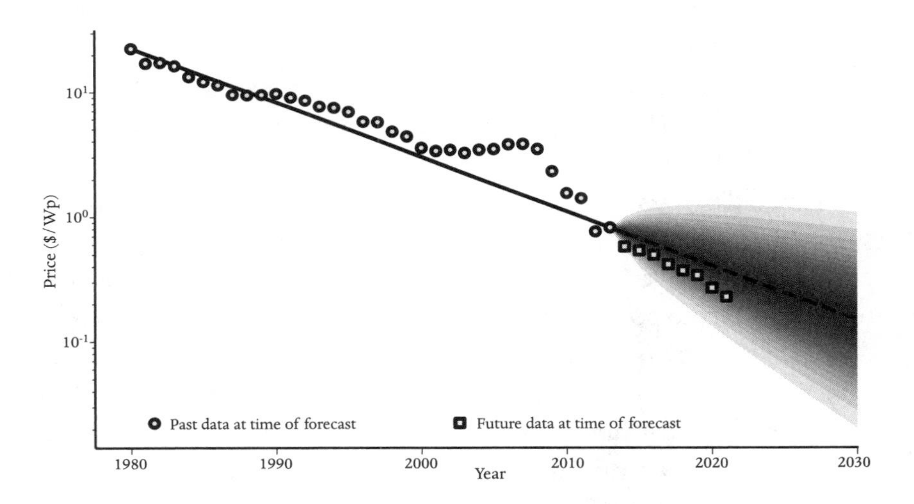

Figure 17. *Prediction of the likely cost of solar-photovoltaic (PV) modules made in 2013.* Prices are in dollars per watt-peak, meaning the price to buy a module that can generate a watt of energy in full sunlight. The grey circles are historical prices of PV modules and the squares are measured values since the prediction was made. Each band corresponds to a probability of 5 per cent; at any given time, there is a 5 per cent probability that the future price will be in the white area outside of the bands.

This method is extremely useful, but it gives no insight into what actually causes technological improvement. With hindsight, it is possible to tell causal stories. For example, the flattening in solar-PV costs between 1999 and 2006 was caused by material shortages, and the plunge immediately afterward was caused by China's entering the market. These kinds of events would have been difficult to forecast, and our method doesn't attempt to do so. Each technology has its story, but our model doesn't need to know anything about those stories. It just takes advantage of the fact that technological trends are highly inertial – once a technology begins a trend, it usually continues that trend. While technologies sometimes break away from their trends (each story is special), our model also takes this into account in estimating the future uncertainty.

While Moore's Law and Wright's Law give similar forecasts for the data in the Performance Curve Database, from a policymaker's point of view they lead to completely different conclusions. Suppose a policymaker wants to encourage the success of a technology through feed-in tariffs that support deployment – for example, by compensating households for putting solar panels on their roofs. Moore's Law implies that costs will go down at a given rate no matter what, so if Moore's Law were literally true, money spent on a feed-in tariff would be wasted. In contrast, Wright's Law suggests that costs will drop only through effort. This brings up a question of cause and effect: Do costs go down because cumulative production goes up (causing effort to go up)? Or does production go up because costs go down on their own, thereby increasing demand? Policymakers need to know the answer to this question to invest our tax dollars wisely.

To test the cause and effect of Wright's Law, François and I tried to find a good natural experiment and settled on World War II.[19] During World War II, the United States went from being the eighteenth-largest military in the world to producing two-thirds of worldwide military equipment (for both sides combined). We know that President Roosevelt didn't say, 'Tanks are getting cheaper, let's build more.' Rather, the US spent whatever it needed to win the war. Furthermore, while production rose rapidly from 1939 until the middle of the war, it declined rapidly as the war ended. Since production stopped increasing exponentially after the war's end, this makes it possible to distinguish between Wright's Law and Moore's Law.

As usual, collecting data was the biggest hurdle. We hired a research assistant, Diana Greenwald, who trolled through the archives in Detroit and Washington, photographed microfiches, sent them to India for transcription and put together a comprehensive data set for hundreds of products produced for the US war effort, ranging from high-tech items like radars to low-tech items like blankets. François performed a careful data analysis, showing that about half of the cost decline for individual technologies is explained by their increase in cumulative production. The other half is explained by the overall trend; the costs of all technologies tended to come down as the war effort ramped up. Thus we can say with some confidence that increasing cumulative production can drive prices down, even if this is not the full story.

So, our answer to policymakers is yes: Subsidies can bring technologies that follow Wright's Law down their learning curves and lower their prices. This doesn't hold for all technologies, because some, like fossil fuels and nuclear power, don't follow Wright's Law. But for those that do, Wright's Law provides useful guidance for finding the cheapest and fastest way to mitigate climate change. (In addition, James McNerney's model in Chapter 3 shows that improvements also come from enhancements to the inputs of a given technology, so, for best results, subsidies need to be distributed thoughtfully.)

A rapid green-energy transition will likely save us money

We applied our forecasting methods to the green-energy transition in a research project led by Rupert Way. Rupert began this project after finishing his PhD in differential geometry, an abstract field of mathematics with no relation to climate change. However, he was passionate about climate change, wanted to do something practical, and convinced me that he could quickly learn what he needed to know. Rupert became an expert on energy technologies, collected an enormous amount of data, and built an energy model with some help from Matt Ives, Penny Mealy and myself.[20]

The history of energy provides some perspective and motivates our model. How has the energy system changed in the past? Figure 18 shows how the energy landscape has evolved over the last 140 years. The plots have a logarithmic scale, in powers of 10, so exponential trends become straight lines. The panel on the left gives the historical costs of the principal energy technologies and the panel on the right gives their deployment. In the left-hand panel, the diagram becomes congested as we approach the present, with most of the curves on top of one another, indicating that we are in a period of unprecedented energy diversity. There are many technologies competing for dominance, all with global average costs near $100 per megawatt-hour.

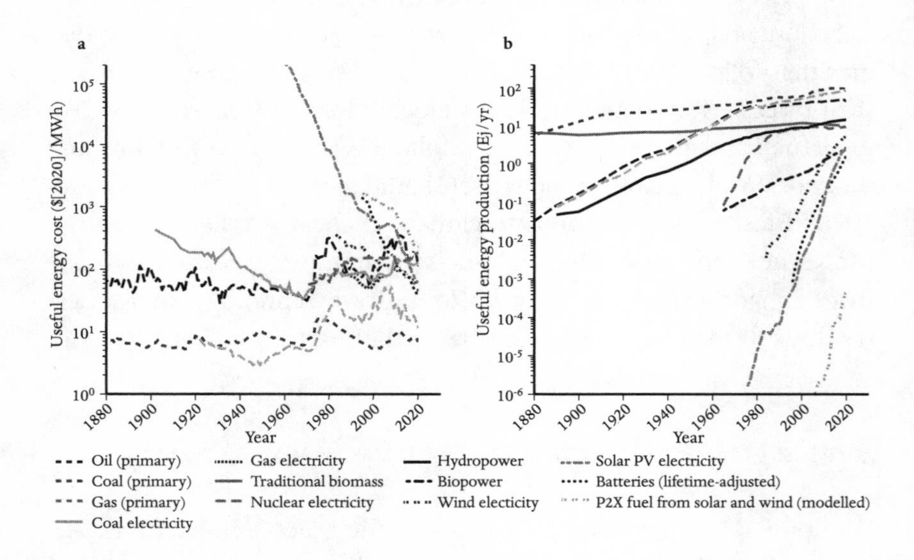

Figure 18. *Historical global cost and deployment of energy technologies*. The panel on the left shows the inflation-adjusted useful energy costs (or prices for oil, coal, and gas), while the panel on the right shows the global production of useful energy.[20]

The long-term trends show how this competition is likely to be resolved. Fossil fuel prices are volatile, but (inflation-adjusted) prices now are similar to what they were 140 years ago.[21] In contrast, for several decades the costs of solar photovoltaics, wind and batteries have dropped (roughly) exponentially, at a rate near 10 per cent per year. If these trends continue, renewables will soon become much cheaper than fossil fuels, even when the costs of energy storage are considered. They are already substantially cheaper than nuclear power, whose cost rose by about a factor of 3 since its first commercial use in 1956 and then reached a plateau. In contrast, the cost of solar cells has decreased by more than 3 orders of magnitude since its first commercial use in the Vanguard satellite in 1958.

The panel on the right shows how the deployment of energy technologies has evolved since 1880. It documents the slow exponential rise in the production of oil and natural gas over a century, until they eventually replaced traditional biomass around 1955 and caught up with coal around 1980. It also shows the rapid rise and plateauing of nuclear energy. But perhaps the most remarkable feature is the dramatic, exponential rise in the deployment of solar PV, wind, batteries and hydrogen electrolyzers over the last decades as they transitioned from niche applications to mass markets. Their annual rate of increase ranges from about 100 per cent for electrolyzers to about 20 per cent for wind. Nuclear energy also underwent a rapid expansion in the 1960s, but, unlike nuclear energy, renewables all have exponentially-decreasing costs. The combination of exponentially-decreasing costs and exponentially-increasing deployment has never occurred for any other energy technologies. This positions renewables to challenge the dominance of fossil fuels within a decade.

Let's pause to take in some good news about what is already underway. Look carefully at the steeply rising curves for solar, wind and batteries in the right panel, which are growing at unprecedented rates. Rapid action on climate change is already in progress, though few people are aware of it. In part, this is because the data is difficult to obtain, but I think it's mostly because our intuition for rapid exponential growth is poor. To see it requires viewing the world through logarithmic glasses. Just when we desperately need energy technologies to replace fossil fuels, they are appearing and growing at an exponential rate! If most of us haven't taken this in, it's because exponential change starts out small, sneaking up on us before suddenly appearing onstage. Assuming that historical trends persist, renewables will become a substantial part of the energy system over the next few years.

We need to solve the problem of intermittency for renewable power sources to fully replace fossil fuels. Solar and wind power are variable, so we need long-term energy storage. Overnight storage can be accomplished by batteries, but they aren't a good solution for longer-term storage. A better solution is provided by hydrogen-based fuels, in which solar or wind electricity is used to make hydrogen, which is then used to make fuels such as ammonia, which are denser and easier to transport and store. Hydrogen-based fuels can then be used for any application where fossil fuels are currently used. This includes making electricity; they can be stored and then cleanly burned later (without carbon emissions) to generate electricity when sunlight or wind are lacking.[22] Although the electrolyzer technologies needed to make hydrogen-based fuels are still expensive, their costs, too, have been declining, and they also show rapid exponential growth (about 100 per cent per year). However, the deployment of hydrogen-based fuels still lags behind the other technologies, and we need to do everything we can to maintain and even accelerate their growth.[23]

But will clean-energy technology costs continue to drop at the same rate in the future? And what does this uncertainty imply for the overall cost of the green-energy transition?

To put my group's answers to these questions in perspective, let's first discuss how mainstream economics models answer them. The dominant approach is via *integrated-assessment models*. The Nordhaus DICE model is a simple example, but there are also much more complicated models that render the energy system in great detail. For a given target to limit global warming, integrated-assessment models calculate the 'optimal' energy transition that maximizes the utility of a representative household. These models are normative, not descriptive. In other words, they attempt to describe the path we should take, rather than the path we're likely to take.

Simple models like the Nordhaus model do everything at an aggregate level, meaning with a single representative agent standing for everyone in the world, and based on global abatement and damage functions, averaging over all technologies and all parts of the world. So many different things are lumped together that it is difficult to compare to reality, and impossible to properly take the heterogeneous nature of technological change into account. Many of these models simply assume that technology costs remain roughly where they are now.

More complicated integrated-assessment models render the energy system in detail and model technological improvement using Wright's Law (though, so far, not with our improved version including error estimates). They then constrain climate emissions following a given scenario and compute the set of decisions that maximize utility in this scenario. Among other things, the output of these models includes the deployment and cost of each energy technology over time.

An alternative is provided by the World Energy Model, developed by the highly influential International Energy Agency (IEA). The IEA, a consortium of thirty-one countries, was founded after the oil crisis of 1973–74 to oversee oil supplies and has since broadened its mandate to investigate related global energy issues, including the climate-induced energy transition. Its World Energy Model is a complicated simulation covering supply, demand and innovation for all of today's significant energy technologies. It makes 'medium to long-term energy projections'. (The word *projections* in effect ducks the responsibility associated with the word *forecast*.)

Integrated-assessment models and the World Energy Model have been in existence since the early 1990s, so we now have almost thirty years of data available to check their track records. To summarize briefly, their performance has been terrible. In our paper, led by Rupert Way, we examined 2,905 projections made by various integrated-assessment models for the annual rate at which solar-PV costs would fall between 2010 and 2020. The mean value of their projected annual cost reductions was 2.6 per cent, and all were less than 6 per cent. In stark contrast, during this period solar-PV costs actually fell by 15 per cent per year. The IEA forecasts are similarly poor (and have been since the inception of their model). Both types of models have consistently and wildly underestimated the deployment of renewables and overestimated their cost. My simple extrapolation of the exponentially decreasing trend made a far better forecast.

Why have these models done so poorly? Integrated-assessment models have typically put in assumptions by hand about the maximum deployment and maximum growth rates for renewables, and they typically impose *floor costs*, corresponding to fixed levels that

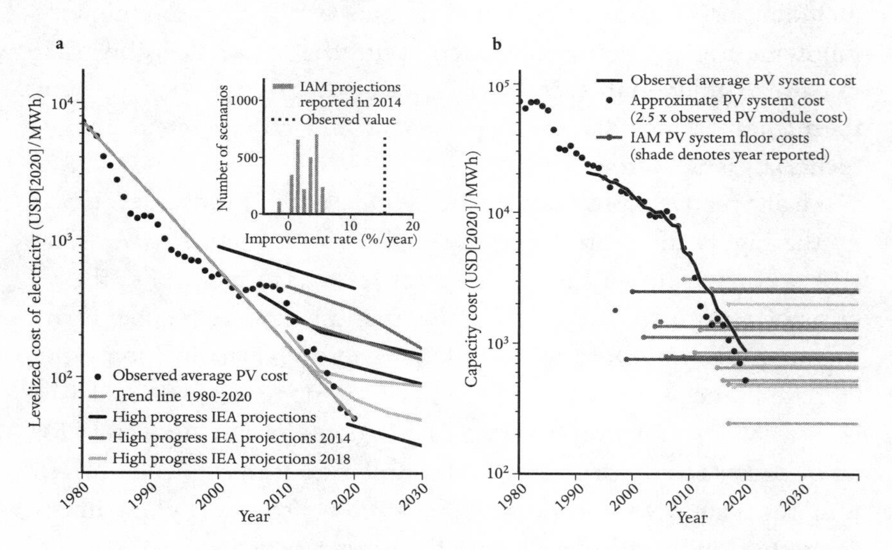

Figure 19. *Track record of projections for PV costs and floor costs.* The left panel compares projections for PV costs to what actually happened. The right panel compares PV costs to floor costs assumed in various integrated-assessment models, demonstrating that PV costs have consistently penetrated assumed floor costs.

costs can never fall below. Figure 19 illustrates how floor costs assumed in the past in integrated-assessment models have repeatedly been violated. There is no empirical evidence supporting them, and the only arguments for imposing them are based on supposed 'common sense'. The reasons for the failure of the IEA's World Energy Model are less obvious – it is a complicated model, and the code is not public – but apparently they rely substantially on expert opinions instead of data extrapolation. Expert opinions have generally been less reliable for predicting technological progress than data extrapolation.[24]

While these models have been updated in an attempt to correct the systematic biases in their past projections, and they are improving, their view of the future continues to differ from what one would infer from historical trends. Even in its Net Zero Emissions by 2050 scenario, the current IEA projection is that solar will reach only 16 per cent of electricity generation in 2050; this was recently updated from its previous projection of 11 per cent, which itself had been updated from a still-lower number. All of the IEA adjustments have been in the same direction but have still not changed sufficiently to match the historical trend. They have always been wrong by a wide margin, underestimating the deployment of renewables and overestimating their cost.

In our 2022 paper, 'Empirically Grounded Technology Forecasts and the Energy Transition',[25] we took a different approach. Unlike the models discussed above, which generate their projections for the deployment of different technologies endogenously, we *assumed* hypothetical scenarios for technology deployment and calculated the probability distribution for their costs using the methods discussed in the previous section. Rather than trying to compute optimal scenarios, we simply examined the cost of three representative scenarios: Fast, Slow and No Transition. Our purpose was not to find the best possible energy transition but to show that if current trends continue for even another decade, we can make the green-energy transition quickly and cheaply.

The deployment of technologies tends to follow an S-curve, in which deployment increases exponentially and then levels off when the technology saturates its market.[26] This is evident, for example, for nuclear power in the right panel of Figure 18. The key difference between our three scenarios is that the historical deployment trends for renewables level off at different times.

In our Fast Transition scenario, renewable energy and storage technologies maintain their current deployment growth rates for a decade, replacing fossil fuels in two decades. The S-curve doesn't level off until renewables become dominant, after which their deployment slows to grow at 2 per cent per year. We chose 2 per cent growth because that's the rate at which global energy usage has been increasing for many decades; we assume that the past trend for energy demand will continue into the future. Short-term storage and electrification of most modes of transportation are achieved with batteries (some in electric vehicles), while long-duration energy storage and all hard-to-electrify applications are served by hydrogen-based fuels. We tried to be conservative, erring on the side of being pessimistic about renewables; for example, we assumed we will need enough stored energy to run the entire global energy system even if the sun stopped shining and the wind stopped blowing at the same time for a month everywhere around the globe (which we estimate is likely more than is actually needed).

For comparison, we also assumed a Slow Transition scenario, in which current rapid-deployment trends for renewables slow down immediately, so that fossil fuels continue to dominate until mid-century. Finally, again for comparison, we assumed a No Transition scenario, in which the energy system remains in its current form and each source of energy grows proportionally at a 2 per cent rate (meaning fossil fuels are locked in forever). This is close to what are now called 'worst case' scenarios (which were until recently called 'business as usual' scenarios).

When we apply our cost-forecasting methods to these three scenarios, the surprising result is that Fast Transition wins by a good margin – many trillions of dollars in savings over the No Transition scenario. To state an exact figure, one must compare future savings to current costs; this is usually done by assuming a discount rate, which puts more weight on the present than the future. The Fast Transition costs a bit more now, but it still produces substantial savings in the future – and for all reasonable discount rates. With the 1.4 per cent discount rate advocated in the Stern Review on the Economics of Climate Change, compiled by the UK government in 2006, the average savings of our Fast Transition would amount to $12 trillion.[27] Because rapid deployment brings us down the learning curve faster, so that costs become lower sooner, the Fast Transition is also substantially cheaper than the Slow Transition.

There are potential barriers to implementing the renewable-energy transition quickly. Perhaps the most important is expanding the capacity of the electrical grid. Electricity plays a much larger role in a world of renewable energy – all vehicles, for example, will need to switch from fossil fuels to either electricity or hydrogen-based fuels. We estimate that grid capacity will need to be expanded by a factor of 4. Although the grid costs for a rapid green-energy transition are substantial, we forecast that they are likely to be more than offset by lower energy costs. In 2050, for example, our estimated global annual expenditure on the electricity network for the Fast Transition is about $670 billion per year, compared with $530 billion per year for the No Transition. However, the expected total system cost in 2050 is about $5.9 trillion per year for the Fast Transition and $6.3 trillion per year for the No Transition. Thus, although the additional $140 billion of grid costs might seem expensive, it is significantly less than the savings that come with cheaper energy.

Another potential barrier is the rapid deployment of technologies to make hydrogen-based fuel. We assume that their current rapid exponential growth continues, but it is vital that this trend not be interrupted.

We also examined a nuclear scenario, in which fossil fuels are replaced mostly by substantially ramping up nuclear power. Nuclear power is already more expensive than renewable technologies, and the gap will almost certainly only widen in the future.[28] The nuclear scenario is by far the most expensive, with an average cost of

$15 trillion more than No Transition and $27 trillion more than the Fast Transition.[29]

In Chapter 1, I cited disagreement over austerity policies as an example of the lack of consensus in economics. The controversy over the cost of the climate transition is an even more urgent example. This is not just a matter of academic debate. High cost estimates have dampened enthusiasm for making the necessary changes quickly. If we are right that mitigating climate change is a net economic benefit, and it is actually cheaper to act quickly, then there are few remaining barriers to making the transformation now, other than the vested interests of the fossil-fuel industry and NIMBY resistance to grid expansion. Climate economics is an area where complexity economics can make a huge difference to the world.[30]

We have explored the virtues of complexity economics and its accomplishments to date, but its fullest potential lies ahead of us. Part V presents my vision of its potential to dramatically improve the reliability of economic models, and hence to help us formulate policies for solving some of the world's biggest problems. What we have accomplished so far only scratches the surface of what is possible.

PART V:
Modeling for a Better Future

We need models that can help navigate the big challenges that we face, problems like climate change, inequality and financial instability. And we need more than just conceptual understanding – what we really want are models that can provide *quantitative* answers, based on testable predictions. This requires building complexity-economics models that are carefully grounded in data and that take all the important effects into account. The following chapter discusses how to model these big challenges, with some reflections about what is needed to make this happen. The envisioned models make full use of state-of-the-art computing and data technology. While ambitious, they are achievable with sufficient investment. The effort involved is substantial, but the potential payoff is massive.

14.

Guidance to Solve Some of Our Big Problems

Keeping the economy on a steady course

As I write, we are in a period of high inflation and everyone is concerned about prices. Wages aren't keeping up. Unemployment remains low, which is a good thing, but it may be fueling inflation. The US Federal Reserve is raising interest rates, trying to do a balancing act to calm inflation without causing unemployment to soar and send us into a recession. The pundits are debating: Is inflation just temporary or will it be persistent? How much do we need to raise interest rates? Are there other policy levers besides interest rates that could achieve these goals more effectively? If so, how should we use them? Controlling inflation requires understanding the interaction of many different factors – the money supply, lending, inequality, supply chains, oil prices and foreign investment, among others. To properly control inflation, we need models that take all these factors into account, in order to understand how they will respond to the policy levers the government can wield.

Inflation is the problem of the moment, but, as we move through the business cycle, the economy keeps presenting us with new problems. Agent-based models will eventually prove a better guide than the mainstream tools currently available, but this requires taking them to a new level. Existing macroeconomic agent-based models – for example, the EURACE model built by Herbert Dawid's group in Bielefeld, the Keynes-Schumpeter model built by Giovanni Dosi's group in Pisa, and the pioneering series of models built by the 'Italian cats' Domenico delli Gatti and Mauro Gallegati – model the economy in considerable detail and have provided useful insight into many macroeconomic questions, but they remain qualitative rather than quantitative tools.[1] They can tackle problems like inflation and the causes of business cycles for a hypothetical economy but not for a specific economy at a specific time. Building on earlier models, Sebastian Poledna and his collaborators have recently developed an agent-based model that was used to make quantitative predictions for the Austrian economy.[2] They tested it head-to-head against the best DSGE models for Austria and showed that its predictions were roughly as good. The Bank of Canada is now using a further-developed version of this model in-house. But as already noted, the performance of DSGE models is poor – we need to do better.[3]

Several groups, including mine, are working to build agent-based models that can out-predict DSGE models. Our new model incorporates several innovative features: For example, rather than using a representative household, it features a demographically accurate synthetic population with millions of households (matching age, education, race and consumption habits). Instead of using a representative firm, we model the behavior of tens of thousands of the largest firms, in one-to-one correspondence with real firms. Instead of modeling a single country, we use the fifty-or-so countries where data is available, making it possible to track economic disruptions as they ricochet around the globe.

Our new model aspires to match all the properties of the economy as accurately as the data allow and is both a microeconomic and macroeconomic model at the same time. It can predict key indicators, like aggregate unemployment, but it can also make micro-predictions, like the effects on a particular industry, or even a particular firm in a given industry. It will allow us to predict distributional properties, like how a given policy will affect inequality or how inequality affects economic output, with far more fidelity than any existing macroeconomic model.

Building models like this is not easy. It requires the acquisition and curation of many large data sets to ground the model and test that it is working properly. It requires the development and maintenance of software libraries to implement models and incrementally improve them. It requires ample cloud-computing resources to run and test models. It requires a large team of researchers who have the necessary skills and domain knowledge to monitor model performance, diagnose failures and correct them. The resources needed to do all this properly are substantial, but as we will discuss in the final chapter, the payoff could be huge.

Reducing inequality

Since the 1970s, economic inequality has grown in many developed countries, particularly in the United States and Great Britain. For half a century, American workers have gotten poorer: US real wages for ordinary (non-supervisory) workers are lower than they were in 1970. Over the same time period, the income of the top 0.1 per cent has grown fivefold, to more than $5,000,000.[4] The good news is that the economic gap between developed and developing countries has diminished, although the disparity remains enormous. Mexico's GDP per capita, for example, is only one-sixth of that of the United States, and the GDP per capita of many African countries is much smaller. Inequalities create friction between the haves and have-nots that eats away at the fabric of society. A certain level of inequality is arguably inevitable, but for our global civilization to function well, and fairly, it needs to be reduced to sensible levels.[5]

As inequality continues to disrupt society, economists like Thomas Piketty and others have shown that government policies can have unintended side-effects that exacerbate the problem.[6] There are overt examples, such as the 2008 financial crisis, when the US government bailed out banks while homeowners with subprime mortgages were left hanging. But there are many other, less obvious examples, like quantitative easing programs, in which central banks buy up bonds or other assets to control interest rates and stimulate the economy. While this might accomplish macroeconomic goals, it can have the undesirable side-effect of increasing inequality (quantitative easing lowers interest rates, which boosts the stock market while doing nothing for the poor). To gain a full understanding of any policy, we need to consider its implications for all the parts of the economy.

Understanding inequality has recently become a hot topic in mainstream economics, but standard theoretical models struggle to provide reliable quantitative answers; they cannot properly incorporate inequality without sacrificing verisimilitude in other ways. HANK models replace the representative household of other DSGE models (see Chapter 5) with a *continuum* of representative households with different earning power, ranging from poor to rich. These models can incorporate effects such as the fact that rich households have a much higher savings rate than poor ones. But, while the HANK models incorporate inequality, they do so only crudely, and they do not include other essential features of the economy, such as the heterogeneity of the production network.

Agent-based models, in contrast, can model inequality effects with little effort. The use of synthetic household populations can capture the full heterogeneity of real households, whose income and wealth can change over time. The ability to accurately predict policy outcomes affecting inequality is one of the biggest strengths of agent-based models.

The use of synthetic households builds on a method called *microsimulation* that has been used for decades. It sits on the edge of the mainstream – the pioneers of the method never received Nobel Prizes, and microsimulation is not considered a sexy topic in economics research – but it's a key tool for policy analysis. Microsimulation models are often used, for example, to evaluate the demographic effects of tax policies. A synthetic population is matched to the real population, and the model then simulates how tax policies affect the balance sheet of each household.

My group and others are borrowing methods from microsimulation in developing our new agent-based models, but with one essential difference: In a microsimulation model, there's no feedback between consumers and producers. If consumers spend less as a result of a tax, these models don't capture the effect this would have on producers, who might cut production and fire workers, thus increasing unemployment, and so on. Agent-based models consistently take the interactions of the various parts of the economy into account. The indirect effects of policies can often be much larger than direct effects.

How does inequality affect growth? Does increasing inequality bring about slower growth, or does it happen the other way around? Or is the relationship between the two context-dependent? There's a tradeoff between the two effects: Growth requires investment, but it also requires consumers. When wealth becomes too concentrated, there is plenty of capital but less consumption. But if wealth becomes too equal, many believe the economy will stall due to lack of investment capital. Inequality also affects interest rates and inflation – concentrating wealth takes money out of circulation, which is deflationary. This may be an important cause of the low inflation and interest rates of the past two decades.

Properly understanding inequality requires good quantitative models, and our new agent-based models should be able to accomplish this.[7] By using synthetic populations, we can precisely understand who will be the winners and who will be the losers from changes in policies. Agent-based modeling offers the possibility of creating a digital laboratory where any proposed new policy can be debugged, and its effect on inequality understood, before it's implemented. This could reduce poverty and unemployment, and save governments from painful mistakes with disastrous effects on the poor.

Avoiding financial crises

Chapters 10 and 11 discussed how monitoring market ecology could help regulators do a better job of keeping the financial system on a steady course. The regulators in almost all developed countries could easily gather the data to monitor the balance sheets for every actor in the system. As my graduate student Maarten Scholl demonstrated in his thesis, it is possible to classify market participants according to their trading strategies.[8] As predicted by my theory of market ecology (see Chapter 10), monitoring the wealth invested in each type of strategy would enable regulators to know when destabilizing trading strategies are becoming large enough to cause problems. The ability to accurately simulate markets, as demonstrated by Aymeric Vie (with help from Maarten), could be immensely valuable. As already mentioned, having such a financial laboratory could allow us to evaluate new financial instruments and avoid financial crises.

While the 2008 financial crisis caused many problems within the financial system, even greater damage was inflicted on the economy as a whole. To understand how disruptions flow back and forth between the financial system and the real economy, we need to merge models of financial crises with macroeconomic models. This requires building detailed maps for how funding flows from the financial system to the real economy, linking the digital laboratory for the financial system with the digital laboratory for the real economy. This isn't difficult to do conceptually, but it requires gathering an enormous amount of data. As before, the task is substantial but the rewards could be far bigger.

Recovering from disasters

Financial disasters have global effects. So do other kinds of disasters: The Fukushima earthquake and tsunami, Hurricane Katrina, the COVID pandemic and the war in Ukraine are only a few recent examples. Even local events, like Fukushima, can have effects that are felt around the world because of their links to the global production network. The economics of disasters are very different from ordinary economic behavior – the economic shocks from disasters tend to be large, localized and sudden. Though each disaster is unique – Zeus is capricious – the responses to them are similar: Disasters cause disturbances in supply and/or demand, producing shortages of specific goods and services, which in turn cause disruptions to production, and so on.

Our model for the impact of the COVID pandemic on the UK economy, discussed in the Prologue, provides a template for modeling the economic consequences of disasters in general. That model allows us to anticipate how the shocks caused by a disaster will propagate through the economy, predict how much damage will be caused and explore strategies for mitigating that damage. Because disasters often directly affect only a few specific industries and geographic locations, aggregate models are of little use – there are many devils in the details that require fine-grained models to understand. Our original model was constructed at the level of fifty-five industries and was based on a few representative households. With sufficient resources, we can create a model at the level of individual firms with a demographically and geographically accurate synthetic population. Such a model would not only be more accurate but would answer many more questions.

Disasters happen quickly and are usually unexpected, leaving little time to build new models. One instead wants a library of models that are ready to go, or can at least be quickly refitted for a new situation. Good disaster models would also allow us to anticipate the likely harm from future disasters, such as more intense hurricanes due to climate change, and plan to minimize their effect.

Reducing climate change

As discussed in Part IV, the changing climate is probably the most urgent problem humanity as a whole has ever faced. While the data-driven techniques described there are useful tools, they address only a small part of the problem. To help guide us through the necessary green-energy and agriculture transition, we need models that can help us plan. How should a country best develop its economy to profit from the transition? What is the right strategy for a business to position itself to take advantage of the new economy? Navigating these questions requires a better understanding of how the transition will play out for the many facets of the economy, including energy, agriculture, innovation, finance, production and labor.

To try to answer these questions, my postdocs, students, ex-students and I are designing what we call the Climate Policy Laboratory. The purpose of the Climate Policy Laboratory is to analyze and assess strategies for mitigating climate change in order to transform the global economy as rapidly, fairly and efficiently as we can. It will allow us to evaluate the effectiveness of the various strategies available to decision-makers at levels ranging from households to firms to nation-states.

Our envisioned Climate Policy Laboratory takes a modular approach. As represented in the schematic diagram below, it's composed of eight modules, each corresponding to a different facet of

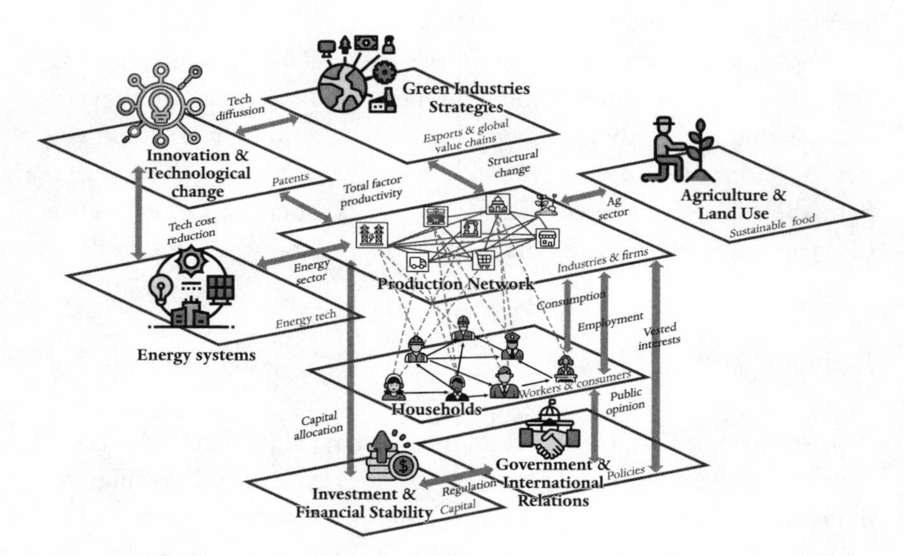

Figure 20. *The Climate Policy Laboratory*. The modules represent the different components of the economy, and can be used separately or in tandem. To keep the diagram comprehensible, we show only a few of the most important connections linking the modules. (Government & International Relations, for example, affects everything.)

the climate-change mitigation problem. These modules can be developed and used by themselves, and tested on their own against real data. They can then be integrated to understand how they interact. For example, the Production Network module affects what happens to industries or firms, and the Households module simulates consumption of their products and the labor supplied to them – each affects the other. When all the modules are used together, they create a fine-grained model for long-range economic growth and political action on a number of scales, which could help guide us through the green-energy transition.

Each module has inputs and outputs that can include physical goods and services and flows of money or information. For most of the modules, outcomes depend on agents, such as government policymakers, business executives or household members. Our goal is to enable decision-makers to understand, in concrete terms, how their actions will affect them and their organizations in the global context of the green-energy transition.

The eight modules of the Climate Policy Laboratory are briefly described below:

- *The Production Network Module* contains the industries and firms that provide goods and services. This module tracks the inputs and outputs of each firm and each industry, and the products they make, through time. Our goal is to do this precisely – that is, the firms will be recognizable companies, realistically headquartered in specific countries. This module allows us to envision possible scenarios for the green transition in great detail, as well as to track carbon emissions comprehensively and accurately.[9]
- The *Energy Systems Module* provides detailed techno-economic predictions about the costs of decarbonizing energy producers and energy-using industries, recommending the technology-investment portfolio likely to be most cost-effective for making the transition.

- The *Agriculture and Land Use Module* allows us to understand the green-food transition and the emergence of sustainable production landscapes that deliver multiple ecosystem services, including carbon sequestration.
- The *Households Module* represents people, who are both consumers and workers, using a demographically realistic synthetic population of individuals. It also incorporates the occupational-labor transition model described in Chapter 4 and allows us to track how the supply of occupational labor will shift in response to changes in the production network on the route to net-zero carbon emissions.
- The *Innovation and Technological Change Module* allows us to understand the innovative trends driving the transition. It will help us determine the best technologies to make a speedy transition to net-zero carbon emissions, and help understand how this process can be accelerated. This module is based on an analysis of patents, historical technology performance and other sources of information about technological change.[10]
- The *Investment and Financial Stability Module* tracks where the financial capital needed for the transition comes from, and suggests ways to prevent physical and transitional risks from destabilizing financial markets and spilling over to the real economy (thereby causing recessions and delaying the transition). This module will identify the providers of capital, their linkages to each other and their linkages to the firms in the real economy.
- The *Green Industries Strategies Module* allows us to recommend developmental pathways that countries should follow to take advantage of the transition.[11]
- The *Government and International Relations Module* enables us to differentiate the decision-making processes concerning climate change mitigation in each nation-state,

based on its system of governance and responsiveness to factors such as economic success, popular opinion and vested interests.

We already have prototypes for most of these modules, but it will require substantial resources to develop them for industrial-strength applications, link them together and turn them into practical tools that can be used on a day-to-day basis.

The perceptive reader will have noticed that the proposed models for addressing the big problems have a great deal in common. All of them depend on properly representing firms and industries, and constructing synthetic household populations. The details of the models are different; understanding the short-term economic effects of a hurricane requires a model substantially different from one that anticipates how occupational labor will change over the next twenty years, but those two models also share many components. This creates an economy of scale – aspects of the models can be modular, and therefore may be bolted together as required. Solving one problem makes solving the next problem easier.

Sustainable growth?

Creating a sustainable economy is the biggest challenge facing civilization. Mitigating climate change is an important step, but it is just one of many problems that need to be solved before the economy can operate sustainably without further degrading the biosphere and destroying itself in the process. I believe we can create a prosperous economy with a small environmental footprint, but we don't understand how to do this. Complexity economics can help guide us.

Tim Jackson and many others say that to have a sustainable economy we need to stop growing, or even have negative growth.[12] While I agree that we urgently need to make the economy sustainable as quickly as possible, and I don't think growth is an important goal for its own sake, I worry it may be counterproductive to pit sustainability against growth. There are too many greedy people, too many vested interests and developing countries are unlikely to embrace policies that leave them behind. Fortunately, I think it is possible to achieve growth and sustainability at the same time.

The green-energy transition provides a good example. While renewable energy does have environmental impacts, these are much smaller than those of fossil fuels. As discussed in Chapter 13, we predict that renewable energy will soon make energy cheaper and cleaner than it has ever been. This will be a boon for everyone, and especially for developing countries. Producing energy will still have some environmental effects, but these will be dramatically reduced.

Growth is traditionally defined in terms of GDP. There are much better indicators of prosperity that could provide superior measuring sticks for well-being, but even GDP suggests that things are improving. GDP has been *dematerializing* at a rate of about 0.7 per cent per year for many decades, meaning that the weight-per-dollar of real (global) GDP has been declining. As it evolves, the economy is becoming more information and service intensive – in other words, industries making products with small material needs are comprising a larger and larger fraction of the economy – making each unit of GDP more sustainable than it was before. The rate is still too slow to counterbalance the global average long-term growth rate of roughly 2 per cent per year, but if we could double the rate of dematerialization it would only take a 30 per cent reduction in the growth rate to bring the two into balance. Of course, sustainability is about more than material balance, but this is another step in the right direction.

Granular complexity economics models can help us achieve sustainability. The sustainability of a product depends on the sustainability of its inputs. Different firms in the same industry can have very different environmental footprints. To measure sustainability we need to assess the environmental footprint of every firm and product for the entire production network and track the flow of unsustainability through the production network. Complexity economics models that operate on this scale could provide guidance to help us coax the economy into becoming sustainable with a minimum of economic pain.

Another effective way to lower our environmental impact is by reducing global population. Several countries, such as Japan, already have declining populations, and most demographers estimate that the global population will begin to decline before the end of this century. Declining populations place a heavier burden on the young to support the old, with side-effects such as declining housing prices, so it brings problems of its own. As already mentioned, complexity economics models can easily be built with demographically accurate synthetic populations, and could help us better understand how to run an economy with an aging population.

If I am right, and we can show that sustainability and growth are compatible, sustainability will be easier to achieve. If I am wrong, we will need to convince people to stop growing the economy and plan carefully to make de-growth as painless as we can. Either way, complexity economics can help us achieve sustainability as soon as possible.

This chapter has presented a sketch of how agent-based models could be used to help guide us through some of our biggest problems, but what do we need to do to realize this goal, and what economic benefits would it provide?

15.

Removing the Roadblocks

I was taught the way of progress was neither swift nor easy.

Marie Curie

Complexity economics is still in its infancy, and a great deal of work remains to be done to realize its full benefits. What are the costs and what would the benefits be? How long will it take?

Collecting better data

Data is the principal driver of scientific progress. There's a bidirectional feedback between data and theory: Data both inspires the development of theories and makes it possible to test them, while theories determine the type of data we collect. Theories can only be proved wrong or right by data; when this doesn't happen, theories become detached from reality, and science gets lost.

Better data is of central importance for improving agent-based models. Unlike traditional mainstream models, which need aggregate data, agent-based models need fine-grained data about individual firms and consumers to realize their potential. This doesn't mean that they necessarily need data about *specific* firms and consumers, which might violate their privacy, but they need data to tell them how typical firms and consumers behave. Unfortunately, though it is feasible to collect fine-grained data, this often does not happen because current data collection systems are geared to provide the aggregate data needed by mainstream models. This creates a chicken-and-egg problem: Because agent-based models are not mainstream, the data are not collected, and because the data aren't available, this handicaps efforts to create agent-based models that make superior predictions, which is a prerequisite for acceptance by the mainstream.

Progress in fundamental physics over the last fifty years illustrates how science succeeds when we have the right data to test theories, and how it is stymied when this is lacking. Fundamental physics made astounding progress during the first two-thirds of the twentieth century, leading to the theory of quantum mechanics, which predicts the behavior of atoms, and the standard model, which predicts the existence and behavior of the elementary particles that are the building blocks of matter and energy. The data engines that drove these discoveries were particle accelerators, which probe the structure of matter and energy by smashing particles together. But making progress requires probing ever-smaller scales, which can only be done by bigger, more powerful accelerators, using exponentially increasing amounts of energy and money. As larger and larger accelerators were built, data flowed in at a steady rate and particle physics advanced rapidly. But each discovery cost more and required more energy than the one before it.

The current state of the art in physics is the Large Hadron Collider at CERN. It is 27 kilometers in circumference and consumes 1.3 terawatt-hours of electricity a year. The electricity alone costs $23.5 million annually, enough to power 300,000 homes. But this is only a small part of its cost – it employs about 2,200 people, with an operating budget of about $1 billion per year. Its crowning achievement to date was the observation of the Higgs boson in 2012, which confirmed theoretical predictions published in 1964. This observation required a forty-year effort and cost about $13 billion.[1]

Gathering the data needed to understand the origins of dark matter and dark energy requires even higher energies. Building an accelerator that could do this would require such a massive expenditure of resources that it's unlikely to ever happen. Absent alternative sources of data, progress in fundamental physics has nearly ground to a halt. The theories are far ahead of the data and thus have become increasingly speculative, and the field is bogged down in doctrinaire battles over the correct approach.

Fortunately, there are no such fundamental limits in economics. Quite the opposite: It is now possible to collect vast quantities of data at a level of detail that would have been inconceivable in 1950. Unfortunately, this isn't happening on the scale needed for agent-based models to fully realize their potential.

In an ideal situation, to render the economy with verisimilitude, agent-based models need data about the behavior of individual households and firms. We don't need information about any actual individuals, but we do need to know in general how individuals make their decisions. For example, what kind of house is a household with a given income and age profile likely to buy? What are they likely to consume? How does this change in a recession? Similarly, we need information about supply chains. In agent-based modeling heaven, we would have access to the invoices and receipts of all the companies in the world. These could then be used to initialize agent-based models capable of simulating the economy in detail and making much better predictions about everything, including how businesses respond to disasters, business cycles and long-term growth.

Since we do not live in agent-based modeling heaven, the complexity-economics group at Oxford and others have been developing methods for reconstructing firm-level production networks. Such data is hard to obtain, because firms don't like to make their invoices and receipts public. There are some countries (Belgium, Chile, Brazil, Ecuador, the Scandinavian nations) where such data exists, but it's hard to access, and most countries don't even collect this data. In the United States, it's possible to obtain partial information about the production network through a combination of public regulatory filings and the media, and there are companies that collect and sell such data. This provides a starting point, but the picture is highly incomplete.

Reconstructing the firm-level network from the available data is like trying to assemble a jigsaw puzzle with hundreds of thousands of pieces, when you only have a small fraction of them and there is no picture on the box. This might sound impossible, but, in a paper led by my clever graduate student Luca Mungo, we showed that machine-learning methods are remarkably effective in solving this problem.[2] We're in the process of developing better methods, ones which combine all possible sources of data on the production network that are good enough for modeling purposes, even if they miss the details.

You may be thinking: Wait a minute! This sounds really intrusive. What about Big Brother? Fortunately, we can do all this without leaking confidential information. New statistical methods have been developed that make it possible to construct randomized synthetic populations that match almost any desired characteristic of a real population, without giving away any information about real individuals. Similarly, there are now companies that specialize in holding the supply chains of individual firms in secure silos, allowing models to probe the silos of many different firms, as needed, without giving away any firm's confidential information. It is vital to preserve confidentiality, and this requires effort and care, but it is feasible to do so.

Developing methods and infrastructure

We need to improve techniques for fitting agent-based models to time series, create an appropriate software infrastructure, and establish standard interfaces and benchmarks so that models can be tested against each other.

Models always have free parameters; these are like knobs that can be turned to produce a whole family of related models. Fitting a model to data requires finding the model within this family that most closely approximates the data.[3] There are now many well-developed procedures for fitting models if the functions they're based on are smooth, meaning that there are no discontinuities or kinks, but the problem becomes harder when this is not the case. Mapping between the inputs and outputs of agent-based models often lacks a simple expression in mathematical terms. Worse still, it's always nonlinear (otherwise no one would bother making an agent-based model). For many years, it has been difficult to fit agent-based models to data in a systematic manner; however, there has recently been substantial progress.[4]

Another significant problem is that agent-based models are written in many different programming languages that are incompatible with each other, making it hard to compare their outputs. There are no standard interfaces that let you mix and match, taking part of model A and using it in model B. There are no benchmark data sets on which to run different models to compare their predictions. Agent-based modeling for economics is still in early stages, but fortunately it is beginning to make rapid progress.

Institutional support

In its present form, complexity economics lacks a home in academia. In the United States, there is only one active research group, led by Robert Axtell at George Mason University. There are perhaps ten small research groups in the rest of the world, and tenure-track positions are extremely rare. The situation in central banks is slightly better: Four or five central banks now support small agent-based modeling efforts. In industry, in contrast, agent-based modelers for problems like traffic control are very much in demand.

The resistance to complexity economics in conventional economics departments is so strong that mainstream academia will likely be the last to accept it. Before this happens, I expect agent-based models to find practical applications in both industry and central banks. This will put pressure on academia, and ultimately the gate to economics departments will open – or, alternatively, new departments doing complexity economics may be created to provide an academic home.

One thing that could dramatically alter the future course of economics is a change in the criteria for the Nobel Prize in economics, which motivates exceptionally ambitious and talented young researchers and strongly influences its scientific agenda. The five original Nobel Prizes – in physics, chemistry, physiology and medicine, peace, and literature – were established in 1896 by a grant from Alfred Nobel. An essential criterion for awarding the prizes in physics, chemistry and physiology and medicine is empirical validation. If a theory has not been shown to be empirically valid, no matter how popular, conceptually appealing or influential it is, no prize is given. This criterion is strict. For example, although the general theory of relativity is perhaps the most profound and remarkable accomplishment in the history of science, Einstein's Nobel Prize was awarded, in 1921, for 'services to Theoretical Physics, and especially for his discovery of the law of the photoelectric effect', because general relativity, while supported by observation of the solar eclipse of 1919, was not considered sufficiently well verified.[5]

The Sveriges Riksbank Prize in Economic Sciences in Memory of Alfred Nobel, colloquially called the Nobel Prize in economics, was established much later, in 1968. Although the nomination and selection processes follow a procedure similar to those used for the other prizes, the selection criteria for economics are closer to those used for the literature and peace prizes than those for the other science prizes. The main criterion is influence within the field of economics. Many of the prizes have been given for theories that have not been empirically verified, and probably never will be, because in hindsight they are very likely wrong. But why shouldn't the same criteria used in other sciences apply to economics as well? If the Nobel Prize in economics made empirical validation an essential criterion, this could prompt a more scientific approach to the study of economics.

For the last decade or so, my colleagues and I have been trying to fund quantitative, large-scale complexity-economics models in the public sector. While we have had some generous supporters (see the Acknowledgements to this book), it is difficult to raise the funding needed to do this at scale. Government funding agencies usually ask the advice of mainstream economists, who do not usually support this idea. With a few exceptions, most private foundations have more conventional charitable goals and do not understand the need.

To bring these ideas to scale, my colleagues and I have started a company called Macrocosm. Our goal is to build large-scale granular models of the economy that make better predictions and reduce them to practice, so they can be used day in and day out to address the important problems discussed above. Our strategy is to fund our efforts through the commercial sector, trying to guide it toward a more sustainable economy, while keeping as much of the modeling power as possible available for public good. Our intention is to put our models in the hands of treasury departments, central banks and other influential actors in the global economy so that they can plan more effectively.

Let's conclude by zooming out to the big picture and thinking about the benefits of complexity economics for humanity as a whole.

16.

Becoming a Conscious Civilization

The principal contradiction in the whole system comes about because of the inability of men to forgo immediate gain for a longtime good [. . .] We do not yet have a sufficient number of people who are ready to make the immediate sacrifice in favor of a long-term investment.

Dwight Eisenhower[1]

If the real economy is the metabolism of civilization, and the financial system its enteric nervous system (its gut brain), then scientific models are an important component of civilization's nascent cerebral cortex. Just as the cerebral cortex enables us to make sense of our environment, detect and avoid dangers, plan ahead and reason, scientific models augment civilization's collective cognitive abilities. We increasingly use science and scientific models to plan for the future and avert environmental threats. As noted, agent-based models provided useful advice during the COVID pandemic. The ability to rapidly create and manufacture vaccines saved many lives. NASA's Scout program, which searches for and performs hazard assessments for potentially threatening near-Earth objects, such as asteroids, may one day save Earth from a major existential catastrophe. Weather forecasting helps us direct air travel and avert major accidents, and climate prediction has alerted us to a major existential global threat, even if we have so far been slow to heed its advice.

As climate change illustrates, we are increasingly altering our environment, and in the course of doing so, creating unintended existential threats to ourselves. While we once thought of nature as an entity unto itself, independent of humanity, no aspect of the environment in today's world is unaffected by human activity. Forty per cent of global photosynthesis is now required to support human civilization. For better or worse, the world has become our garden, and we need to do a better job of tending it.

To return to a biological analogy, in addition to the functions listed above, our cerebral cortex provides us with the ability to understand and model ourselves, which we call *consciousness*. Self-awareness is essential for allowing individuals to cooperate and function as a society. Similarly, as we increasingly dominate the planet, we need better models of our collective behavior in order to survive and prosper. Civilization needs to become more conscious.

Economics plays a central role in creating our collective environment. Agent-based models of the global economy could potentially allow us to better understand how our actions will shape our self-constructed environment and shape the future. This would endow civilization with greater self-understanding and awareness and make a valuable contribution to increasing civilization's consciousness.

Throughout most of history, we have simply let events unfold. History might be a guide, but its lessons can be murky and hard to follow. In economics, statistical models and AI are useful, but they are inherently limited – partly because past events are never enough like events in the present, and partly because the world is always changing. As culture and technology co-evolve, we create new technologies that lead to genuinely new situations, with side-effects that are difficult to anticipate.

We now have the capability to build realistic agent-based models that will allow us to safely experiment with diverse ideas about the best way to run our economy and navigate societal challenges. These models will be pale reflections of the real world, but, by grounding them in empirical data and continually testing them, we can create and experiment with artificial economies that yield useful insights and advice. Even if far from perfect, their practical guidance could allow us to avert a great deal of economic and social pain.

To put this in context, global GDP is on the order of $80 trillion per year. The 2008 financial crisis cost the US alone approximately $10 trillion. Suppose that by spending $100 million over the next five years we could build a well-calibrated agent-based model of the global economy and avert the next financial crisis of similar size. The payoff would be 100,000 to 1, which corresponds to a financial return of 10 million per cent!

But let's make a more sober estimate. Averting the next crisis with absolute certainty is too optimistic. Suppose that the odds of succeeding are only 1 in 100, and that by using a successful model we reduce the impact of the crisis by only 1 per cent, meaning it is only slightly less bad – 99 per cent of what it would have been otherwise. This might not sound like much of an improvement – indeed, I think it is far too pessimistic – but even so the return on the investment would still be 10 to 1, corresponding to a financial return of 1,000 per cent.

Similar payoffs can be reaped by investing in models for other big problems, such as navigating the green-energy transition, which will easily cost us tens of trillions of dollars if we get it wrong, not to mention the potential loss of biodiversity and the danger of increasing inequality and social unrest. There are synergies between these problems; building a modular laboratory to solve the climate problem makes it easier to solve others. State-of-the-art agent-based models of the production network, household consumption and the occupational network will let us do economics on timescales ranging from days to decades. The calculation above suggests that their construction is likely to be a very valuable investment.

That said, it would be misleading to evaluate the benefits of prescient models in purely monetary terms. A financial crisis that costs $10 trillion is not just about rich people losing a lot of money; it entails widespread unemployment, impoverishment, opioid use and all the other problems that arise when people are put under financial stress.

Some might say that trying to smooth out the swings in the business cycle is a bad idea. In Joseph Schumpeter's theory of creative destruction, mentioned in Chapter 2, business cycles and crises serve a useful role because they clean out the dead wood, removing the incumbents to make room for more dynamic, higher-quality businesses. But, even if Schumpeter's theory is true, recessions and crises may not be the only way to achieve creative destruction. Building models like our Climate Policy Laboratory, described in Chapter 14, might help persuade governments to pursue other policies that could keep the economy on a steadier course while allowing for some creative destruction.

In Chapter 14, I outlined how we could build a suite of new models to help us solve some of the biggest problems facing civilization. Doing this right would take a budget of tens of millions of dollars a year. This might sound expensive, but to put it in perspective, we currently spend more than $1 billion a year to smash elementary particles into each other, about $5 billion a year to predict the weather and about $500 million a year on polar research. While I am a fan of all these activities – I am not advocating any cuts in their funding – these figures suggest that it should be feasible to spend a comparably modest amount to explore a new way of doing economics that could help save civilization.

I believe that we can regulate the financial system to do its job well without stirring up volatility and exacerbating inequality. I believe we can implement policies that more fairly distribute profits while increasing our economic output and shared well-being. I believe we can create a sustainable economy that can promote the health of the earth's ecosystems while allowing us to prosper. I believe we can have the best of both worlds, creating a smooth-running economy while allowing enough turnover for the forces of creative destruction to do their job. But to create a road-map for how to achieve these goals, we need to invest the collective resources and effort to build large-scale quantitative agent-based models and let them prove their full worth.

Sometime in the future, we are sure to create detailed agent-based models of the global economy. Such models will be coupled to physical models of the earth's environment, and sociological and political agent-based models of our collective behavior.[2] I have no doubt that this will eventually happen: The question is, when will it happen? We now have the capability – the sooner we do it, the sooner we will reap the benefits.

A crash program to build full-scale complexity-economics models is a good bet: One that could guide us toward greater prosperity, make our planet healthier and help humanity thrive.

Bibliography

Aleta, Alberto, David Martín-Corral, Michiel A. Bakker, Ana Pastore y Piontti, Marco Ajelli, Maria Litvinova, Matteo Chinazzi, Natalie E. Dean, M. Elizabeth Halloran, Ira M. Longini Jr, Alex Pentland, Alessandro Vespignani, Yamir Moreno and Esteban Moro. 2021. 'Quantifying the Importance and Location of SARS-CoV-2 Transmission Events in Large Metropolitan Areas'. medRxiv 2020.12.15.20248273; doi: 10.1101/2020.12.15.20248273.

Alleman, T. W., K. Schoors and J. M. Baetens. 2023. 'Validating a Dynamic Input-Output Model for the Propagation of Supply and Demand Shocks During the COVID-19 Pandemic in Belgium'. https://arxiv.org/pdf/2305.16377.pdf

Anderson, Philip W. 2000. 'Emergence, reductionism, and the seamless web: When and why is science right?' *Current Science*, vol. 78, No. 6.

Anderson, Philip W., Kenneth J. Arrow and David Pines (eds). 1988. *The Economy as an Evolving Complex System*. New York: CRC Press.

Arnold, Robert, Jeanne De Sa, Tim Gronniger, Allison Percy and Julie Somers. 2006. 'A Potential Influenza Pandemic: Possible Macroeconomic Effects and Policy Issues', revised July 27, 2006, Congressional Budget Office. https://www.cbo.gov/sites/default/files/109th-congress-2005-2006/reports/12-08-birdflu.pdf

Arthur, W. Brian. 1991. 'Designing Economic Agents that Act Like Human Agents: A Behavioral Approach to Bounded Rationality'. *American Economic Review* 81 (2): 353–359.

Ibid. 1994a. *Increasing Returns and Path Dependence in the Economy*. Ann Arbor: University of Michigan Press.

Ibid. 1994b. 'Inductive Reasoning and Bounded Rationality'. *American Economic Review* 84 (2): 406–411.

Ibid. 1999. 'Complexity and the Economy'. *Science* 284.5411: 107–109.

Ibid. 2021. 'Foundations of Complexity Economics'. *Nature Reviews Physics*, 3(2): 136–145.

Arthur, W. Brian, John H. Holland, Blake LeBaron, Richard G. Palmer and Paul Tayler. 1996. 'Asset Pricing under Endogenous Expectations in an Artificial Stock Market'. SFI working paper 1996-12-093, https://santafe.edu/research/results/working-papers/asset-pricing-under-endogenous-expectations-in-an-.

Asano, Yuki M., Jakob J. Kolb, Jobst Heitzig and J. Doyne Farmer. 2021. 'Emergent Inequality and Business Cycles in a Simple Behavioral Macroeconomic Model'. *Proceedings of the National Academy of Sciences* 118 (27): e2025721118, doi: 10.1073/pnas.2025721118.

Athey, S. and G. W. Imbens. 2019. 'Machine Learning Methods that Economists Should Know About'. *Annual Review of Economics* 11: 685–725.

Axtell, Robert L., J. Doyne Farmer, John Geanakpolos, Peter Howitt, Ernesto Carrella, Ben Conlee, Jon Goldstein, Matthew Hendrey, Philip Kalikman, David Masad, Nathan Palmer and Chun-Yi Yang. 2014. 'An Agent-Based Model of the Housing Market Bubble in Metropolitan Washington D.C.', https://ssrn.com/abstract=4710928

Axtell, Robert L. and J. Doyne Farmer. 2022. 'Agent-Based Modeling in Economics and Finance: Past, Present, and Future'. *Journal of Economic Literature*, forthcoming, doi: 10.1257/jel.20221319.

Axtell, Robert L. 2024. 'Simon Said: Herb's Research Rules and Life Lessons', to appear in Gigerenzer et al. (eds), *A Companion to Herb Simon*. Edward Elgar Publishing.

Aymanns, Christoph and J. Doyne Farmer. 2015. 'Dynamics of the Leverage Cycle'. *Journal of Economic Dynamics and Control* 50: 155–179, doi: 10.1016/j.jedc.2014.09.015.

Aymanns, Christoph, Fabio Caccioli, J. Doyne Farmer and Vincent Tan. 2016. 'Taming the Basel Leverage Cycle'. *Journal of Financial Stability* 27: 263–277, doi: 10.1016/j.jfs.2016.02.004.

Aymanns, Christoph, J. Doyne Farmer, Alissa M. Kleinnijenhuis and Thom Wetzer. 2018. 'Models Of Financial Stability And Their Application In Stress Tests', in Cars Hommes and Blake LeBaron (eds), *Handbook of Computational Finance* 4: 329–391.

Bachelier, Louis. 1900. 'Théorie de la Spéculation'. PhD thesis, University of Paris.

Bagley, R. J. and J. D. Farmer. 1991. 'Spontaneous Emergence of a Metabolism', in C. Langton, C. Taylor, J. D. Farmer and S. Rasmussen (eds), *Artificial Life II*, proc. 93–140. Santa Fe Institute Studies in the Sciences of Complexity. Redwood City: Addison Wesley.

Bak, Per, Maya Paczuski and Martin Shubik. 1997. 'Price Variations in a Stock Market with Many Agents'. *Physica A* 246 (3–4): 430–453, doi: 10.1016/S0378-4371(97)00401-9.

Baptista, Rafa, J. Doyne Farmer, Marc Hinterschweiger, Katie Low, Daniel Tang and Arzu Uluc. 2016. 'Macroprudential Policy in an Agent-based Model of the Housing Market'. Bank of England Staff Working Paper No. 619, https://www.inet.ox.ac.uk/files/swp619.pdf

Baqaee, David Rezza and Emmanuel Farhi. 2019. 'The Macroeconomic Impact of Microeconomic Shocks: Beyond Hulten's Theorem'. *Econometrica* 87 (4): 1155–1203, doi: 10.3982/ECTA15202.

Bass, Thomas A. 1985. *The Eudaemonic Pie*. Boston: Houghton Mifflin.

Battiston, Stefano, Michelangelo Puliga, Rahul Kaushik, Paolo Tasca and Guido Caldarelli. 2012. 'DebtRank: Too Central to Fail? Financial Networks, the FED and Systemic Risk'. *Scientific Reports* 2: 541, doi: 10.1038/srep00541.

Bauer, Peter, Alan Thorpe and Gilbert Brunet. 2015. 'The Quiet Revolution of Numerical Weather Prediction'. *Nature* 525: 47–55, doi: 10.1038/nature14956.

Baumol, William J. 2012. *The Cost Disease: Why Computers Get Cheaper and Health Care Doesn't*. New Haven: Yale University Press.

Beaudry, Paul, Dana Galizia and Franck Portier. 2015. 'Reviving the Limit Cycle View of Macroeconomic Fluctuations'. NBER working paper 21241. https://www.nber.org/papers/w21241

Beaudry, Paul, Dana Galizia and Franck Portier. 2016. 'Putting the Cycle Back into Business Cycle Analysis'. NBER working paper 22825. https://www.nber.org/papers/w22825

Bednar, J. and S. Page. 2007. 'Can Game(s) Theory Explain Culture?: The Emergence of Cultural Behavior Within Multiple Games'. *Rationality and Society* 19 (1): 65–97. https://doi.org/10.1177/1043463107075108

Beinhocker, Eric D. 2007. *The Origin of Wealth: Evolution, Complexity, and the Radical Remaking of Economics*. Boston: Harvard Business School Press.

Benhabib, Jess, and Kazuo Nishimura. 1979. 'The Hopf Bifurcation and Existence and Stability of Closed Orbits in Multisector Models of Optimal Economic Growth'. *Journal of Economic Theory* 21: 421–444, doi: 10.1016/0022-0531(79)90050-4.

Benhabib, Jess, and Richard H. Day. 1981. 'Rational Choice and Erratic Behaviour'. *Review of Economic Studies* 48 (3): 459–471, doi: 10.2307/2297158.

Bergmann, Barbara R. 1980. 'A Short Description of a Micro-Simulated Macro-Economic Model: The Transactions Model of the U.S. Economy', in B. R. Bergmann, G. Eliasson and G. H. Orcutt (eds) *Micro Simulation – Models, Methods and Applications*. Stockholm: The Industrial Institute for Economic and Social Research.

Black, Fischer and Myron Scholes. 1973. 'The Pricing of Options and Corporate Liabilities'. *Journal of Political Economy* 81 (3): 637–654.

Black, Fischer. 1986. 'Noise'. *Journal of Finance* 41 (3): 528–543, doi: 10.1111/j.1540-6261.1986.tb04513.x.

Blume, Larry and David Easley. 1991. 'Evolution and Market Behavior'. *Journal of Economic Theory* 58 (1): 9–40.

Boldrin, Michele and Luigi Montrucchio. 1986. 'On the Indeterminacy of Capital Accumulation Paths'. *Journal of Economic Theory* 40 (1): 26–39, doi: 10.1016/0022-0531(86)90005-0.

Bookstaber, Richard. 2019. *The End of Theory: Financial Crises, the Failure of Economics, and the Sweep of Human Interaction*. Princeton: Princeton University Press.

Bouchaud, Jean-Phillippe. 2008. 'Economics Needs a Scientific Revolution'. *Nature*, 455 1181.

Bouchaud, Jean-Philippe, Julius Bonart, Jonathan Donier and Martin Gould. 2018. *Trades, Quotes and Prices: Financial Markets Under the Microscope*. Cambridge: Cambridge University Press.

Brock, William A., David A. Hsieh and Blake LeBaron. 1991. *Nonlinear Dynamics, Chaos, and Instability*. Cambridge, MA: MIT Press.

Brock, William A. and Cars H. Hommes. 1997. 'A Rational Route to Randomness'. *Econometrica* 65 (5): 1059–1095, doi: 10.2307/2171879.

Brunnermeier, Markus K. and Yuliy Sannikov. 2014. 'A Macroeconomic Model with a Financial Sector'. *American Economic Review* 104 (2): 379–421.

Buda, Gergely, Vasco M. Carvalho, Stephen Hansen, Alvaro Ortiz, Tomasa Rodrigo, José V. Rodriguez Mora. 2022. 'National Accounts in a World of Naturally Occurring Data: A Proof of Concept for Consumption'. https://sekhansen.github.io/pdf_files/consumption_measure.pdf

Cabannes, Théopile, Marco Antonio Sangiovanni Vincentelli, Alexander Sundt, Hippolyte Signargout, Emily Porter, Vincent Fighiera, Juliette Ugirumurera and Alexandre M. Bayen. 2018. 'The Impact of GPS-Enabled Shortest Path Routing on Mobility: A Game Theoretic Approach'. Transportation Research Board 97th Annual Meeting. https://trid.trb.org/view/1495267

Caccioli, Fabio, Jean-Philippe Bouchaud and J. Doyne Farmer. 2012. 'Impact-Adjusted Valuation and the Criticality of Leverage'. *Risk*, December, 74–77.

Calvet, Laurent and Adlai Fisher. 2001. 'Forecasting Multifractal Volatility'. *Journal of Econometrics* 105 (1): 27–58.

Calvo, Guillermo A. 1983. 'Staggered Prices in a Utility-Maximizing Framework'. *Journal of Monetary Economics* 12 (3): 383–398, doi: 10.1016/0304-3932(83)90060-0.

Carlin, Wendy and David Soskice. 2015. *Macroeconomics: Institutions, Instability and the Financial System*. Oxford: Oxford University Press.

Carlson, Mark. 2007. 'A Brief History of the 1987 Stock Market Crash with a Discussion of the Federal Reserve Response'. Federal Reserve Staff Working Paper 2007–13, Finance and Economics Discussion Series. https://www.federalreserve.gov/pubs/feds/2007/200713/200713pap.pdf

Carro, Adrian and Patricia Stupariu. 2022. 'Uncertainty, Non-Linear Contagion and the Credit Quality Channel: An Application to the Spanish Interbank Market'. Banco de España documentos de trabajo no. 2212. https://www.bde.es/f/webbde/SES/Secciones/Publicaciones/PublicacionesSeriadas/DocumentosTrabajo/22/Files/dt2212e.pdf

Carro, Adrian, Marc Hinterschweiger, Arzu Uluc and J. Doyne Farmer. 2022. 'Heterogeneous Effects and Spillovers of Macroprudential Policy in an Agent-Based Model of the UK Housing Market'. Bank of England Staff Working Paper no. 976. https://www.bankofengland.co.uk/working-paper/2022/heterogeneous-effects-and-spillovers-of-macroprudential-policy-in-model-of-uk-housing-market

Carroll, Christopher D. 1997. 'Buffer-Stock Saving and the Life Cycle/Permanent Income Hypothesis'. *Quarterly Journal of Economics* 112 (1): 1–55.

Cass, David. 1965. 'Optimum Growth in an Aggregative Model of Capital Accumulation'. *Review of Economic Studies* 32 (3): 233–240, doi:10.2307/2295827.

Catapano, Gennaro, Francesco Franceschi, Michele Loberto and Valentina Michelangeli. 2021. 'Macroprudential Policy Analysis via an Agent Based Model of the Real Estate Sector'. Banca D'Italia temi de discussione number 1338. https://www.bancaditalia.it/pubblicazioni/temi-discussione/2021/2021-1338/en_tema_1338.pdf

Charney, J. G., R. Fjörtoft and J. von Neumann. 1950. 'Numerical Integration of the Barotropic Vorticity Equation'. *Tellus* 2: 237–254, doi: 10.1111/j.2153-3490.1950.tb00336.x.

Cherkashin, Dmitriy, J. Doyne Farmer and Seth Lloyd. 2009. 'The Reality Game'. *Journal of Economic Dynamics and Control* 33 (5): 1091–1105, doi: 10.1016/j.jedc.2009.02.002.

Chevalier, J. and A. Goolsebee. 2009. 'Are Durable Goods Consumers Forward-Looking? Evidence from College Textbooks'. *Quarterly Journal of Economics* 124 (4): 1853–1884.

Chinazzi, Matteo, Jessica T. Davis, Marco Ajelli, Corrado Gioannini, Maria Litvinova, Stefano Merler, Ana Pastore y Piontti, Kunpeng Mu, Luca Rossi, Kaiyuan Sun, Cécile Viboud, Xinyue Xiong, Hongjie Yu, M Elizabeth Halloran, Ira M Longini Jr and Alessandro Vespignani. 2020. 'The Effect of Travel Restrictions on the Spread of the 2019 Novel Coronavirus (COVID-19) Outbreak'. *Science* 368 (6489).

Clark, Mills Gardner. 1956. *The Economics of Soviet Steel*. Cambridge: Harvard University Press, doi: 10.4159/harvard.9780674494268.

Coeure, Benoit. 2014. 'What's the Right Size for the Financial Sector?' *World Economic Forum*, 9 September 2014. https://www.weforum.org/agenda/2014/09/benoit-coeure-ecb-financial-sector/

Cohen, Joshua (ed.). 2019. *Economics After Neoliberalism* Cambridge: MIT Press.

Cokayne, Graeme. 2019. 'The Effects of Macroprudential Policies on House Price Cycles in an Agent-based Model of the Danish Housing Market'. Danmarks Nationalbank working paper no. 138. https://

www.nationalbanken.dk/da/publikationer/Documents/2019/05/Working%20Paper_nr.%20138.pdf

Cutler, David M., James M. Poterba and Lawrence H. Summers. 1989. 'What Moves Stock Prices?' *Journal of Portfolio Management* 15 (3): 4–12, doi: 10.3905/jpm.1989.409212.

Cutler, D. M., Poterba, J. M. and Summers, L. H. 1991. 'Speculative Dynamics'. *Review of Economic Studies* 58 (3): 529–546.

Daniels, Marcus G., J. Doyne Farmer, László Gillemot, Giulia Iori and D. Eric Smith. 2003. 'Quantitative Model of Price Diffusion and Market Friction Based on Trading as a Mechanistic Random Process'. *Physical Review Letters* 90(10): 108102–108104, doi: 10.1103/PhysRevLett.90.108102.

Danielsson, J., P. Embrechts, C. A. E. Goodhart, C. Keating, F. Muennich, O. Renault and H. S. Shin. 2001. 'An Academic Response to Basel II'. http://www.bis.org/bcbs/ca/fmg.pdf

Darley, Vince and Alexander V. Outkin. 2007. *A NASDAQ Market Simulation: Insights on a Major Market from the Science of Complex Adaptive Systems*. Hackensack: World Scientific.

Davis, Jessica T., Matteo Chinazzi, Nicola Perra, Kunpeng Mu, Ana Pastore y Piontti, Marco Ajelli, Natalie E. Dean, Corrado Gioannini, Maria Litvinova, Stefano Merler, Luca Rossi, Kaiyuan Sun, Xinyue Xiong, Ira M. Longini Jr, M. Elizabeth Halloran, Cécile Viboud and Alessandro Vespignani. 2021. 'Cryptic Transmission of SARS-CoV-2 and the First COVID-19 Wave'. *Nature* 600: 127–132, doi: 10.1038/s41586-021-04130-w.

Dawid, Herbert, Simon Gemkow, Philipp Harting, Sander van der Hoog and Michael Neugart. 2012. 'The Eurace@Unibi Model: An Agent-Based Macroeconomic Model for Economic Policy Analysis'. Bielefeld Working Papers in Economics and Management No. 05-2012, doi: 10.4119/unibi/2622068.

Day, Richard H. 1981. 'Emergence of Chaos from Neoclassical Growth'. *Geographical Analysis* 13: 315–327, doi: 10.1111/j.1538-4632.1981.tb00741.x.

Ibid. 1982. 'Irregular Growth Cycles'. *American Economic Review* 72 (3): 406–414.

Del Negro, Marco, Gauti Eggertsson, Andrea Ferrero and Nobuhiro Kiyotaki. 2017. 'The Great Escape: A Quantitative Evaluation of the Fed's Liquidity Facility'. *American Economic Review* 107 (3): 824–57.

del Rio-Chanona, R. M., P. Mealy, A. Pichler, F. Lafond and J . D. Farmer. 2020. 'Supply and Demand Shocks in the COVID-19 Pandemic: An Industry and Occupation Perspective'. *Oxford Review of Economic Policy*, graa033, https://doi.org/10.1093/oxrep/graa033.

del Rio-Chanona, R. Maria, Penny Mealy, M. Beguerisse-Diaz, François Lafond and J. Doyne Farmer. 2021. 'Occupational Mobility and Automation: A Data-Driven Network Model'. *Journal of the Royal Society Interface*, doi: 10.1098/rsif.2020.0898.

delli Gatti, D., S. Desiderio, E. Gaffeo, P. Cirillo and M. Gallegati. 2011. *Macroeconomics from the Bottom-Up* (Vol. 1). Springer Science & Business Media.

DeLong, J. B., A. Shleifer, L. H. Summers and R. J. Waldmann. 1990. 'Positive Feedback Investment Strategies and Destabilizing Rational Speculation'. *Journal of Finance* 45 (2): 379–395.

Dessertaine, T., J. Moran, M. Benzaquen and J. P. Bouchaud. 2022. 'Out-of-Equilibrium Dynamics and Excess Volatility in Firm Networks'. *Journal of Economic Dynamics and Control* 138: 104362.

Diamond, D. W. and P. H. Dybvig. 1983. 'Bank Runs, Deposit Insurance, and Liquidity'. *Journal of Political Economy* 91 (3): 401–419.

Diamond, Peter and Drew Fudenberg. 1989. 'Rational Expectations Business Cycles in Search Equilibrium'. *Journal of Political Economy* 97 (3): 606–619, doi: 10.1086/261618.

Diederich, Sigurd and Manfred Opper. 1989. 'Replicators with Random Interactions: A Solvable Model'. *Physical Review A* 39 (8): 4333–4336, doi: 10.1103/physreva.39.4333.

Dosi, G. and R. R. Nelson. 1994. 'An Introduction to Evolutionary Theories in Economics'. *Journal of Evolutionary Economics* 4: 153–172.

Dosi, G., G. Fagiolo and A. Roventini. 2010. 'Schumpeter Meeting Keynes: A Policy-Friendly Model of Endogenous Growth and Business Cycles'. *Journal of Economic Dynamics and Control* 34 (9): 1748–1767.

Duan, Wei, Zongchen Fan, Peng Zhang, Gang Guo and Xiaogang Qiu. 2015. 'Mathematical and Computational Approaches to Epidemic Modeling: A Comprehensive Review'. *Frontiers of Computer Science* 9: 806–826, doi: 10.1007/s11704-014-3369-2.

The Economist, 'Sun, Wind and Drain'. July 29, 2014. https://www.economist.com/finance-and-economics/2014/07/29/sun-wind-and-drain

Edge, Rochelle M. and Refet S. Gürkaynak. January 18, 2011. 'How Useful are Estimated DSGE Model Forecasts?' FEDS Working Paper No. 2011–11. https://ssrn.com/abstract=1810075 or http://dx.doi.org/10.2139/ssrn.1810075

Edwards, Paul N. 2010. *A Vast Machine: Computer Models, Climate Data and the Politics of Global Warming*. Cambridge, MA: MIT Press.

Epstein, J.M. and R. Axtell. 1996. *Growing Artificial Societies: Social Science from the Bottom Up*. Brookings Institution Press.

Epstein, J.M. 2013. Agent_Zero. Princeton, NJ: Princeton University Press.

Erev, Ido, Eyal Ert, Ori Plonsky, Doron Cohen and Oded Cohen. 2017. 'From Anomalies to Forecasts: Toward a Descriptive Model of Decisions under Risk, under Ambiguity, and from Experience'. *Psychological Review* 124 (4): 369–409, doi: 10.1037/rev0000062.

Fama, Eugene F. 1976. *Foundations of Finance*. New York: Basic Books.

Fama, Eugene F. 1991. 'Efficient Capital Markets: II'. *Journal of Finance* 46: 1575–1617, doi: 10.1111/j.1540-6261.1991.tb04636.x.

Fama, Eugene F. and Kenneth R. French. 1996. 'Multifactor Explanations of Asset Pricing Anomalies'. *Journal of Finance* 51 (1): 55–84.

Farmer, J. Doyne (ed.). 1986. *Evolution, Games, and Learning: Models for Adaptation in Machines and Nature: Proceedings of the Fifth Annual International Conference of the Center for Nonlinear Studies, Los Alamos, NM 87545, USA, May 20–24, 1985*. New York: North-Holland.

Farmer, J. Doyne. 2001. 'Toward Agent-Based Models for Investment' in R. Max Darnell (ed.). *Developments in Quantitative Investment Models. AIMR Conference Proceedings*. Boston: AIMR.

Ibid. 2002. 'Market Force, Ecology, and Evolution'. *Industrial and Corporate Change* 11 (5): 895–953, doi: 10.1093/icc/11.5.895.

Ibid. 2004. 'The Everyday Practice of Physics in Silver City, New Mexico', in John Brockman (ed), *Curious Minds: How a Child Becomes a Scientist*. New York: Pantheon Books.

Ibid. 2005. 'The Evolution of Adventure in Literature and Life', presented on November 11, 2005, at a conference organized by Marget Cohen, titled 'Adventure', at Stanford's Center of the Study of the Novel. https://www.doynefarmer.com/adventures

Ibid. 2013. 'Hypotheses Non Fingo: Problems with The Scientific Method in Economics'. *Journal of Economic Methodology* 20 (4): 377–385, doi: 10.1080/1350178X.2013.859408.

Farmer, J. Doyne, Tommaso Toffoli and Stephen Wolfram (eds). 1984. *Cellular Automata: Proceedings of an Interdisciplinary Workshop, Los Alamos, New Mexico 87545, USA, March 7–11, 1983*. New York: North-Holland.

Farmer, J. D., Norman H. Packard and Alan S. Perelson. 1986. 'The Immune System, Adaptation, and Machine Learning'. *Physica D: Nonlinear Phenomena* 22 (1–3): 187–204, doi: 10.1016/0167-2789(86)90240-X.

Farmer, J. D., S. Kauffman and N. H. Packard. 1986. 'Autocatalytic Replication of Polymers'. *Physica D* 22 (1–3): 50–67.

Farmer, J. Doyne and John J. Sidorowich. 1987. 'Predicting Chaotic Time Series'. *Physical Review Letters* 59 (8): 845, doi: 10.1103/PhysRevLett.59.845.

Farmer, J. D. and A. W. Lo. 1999. 'Frontiers of Finance: Evolution and Efficient Markets'. *Proceedings of the National Academy of Sciences* 96 (18): 9991–9992.

Farmer, J. Doyne and Shareen Joshi. 2002. 'The Price Dynamics of Common Trading Strategies'. *Journal of Economic Behavior and Organization* 49 (2): 149–171, doi: 10.1016/S0167-2681(02)00065-3.

Farmer, J. Doyne, Paolo Patelli and Ilija Zovko. 2005. 'The Predictive Power of Zero Intelligence in Financial Markets'. *Proceedings of the National Academy of Sciences USA* 102 (6): 2254–2259, doi: 10.1073/pnas.0409157102.

Farmer, J. Doyne and John Geanakoplos. 2009. 'The Vices and Virtues of Equilibrium and the Future of Financial Economics'. *Complexity* 14 (3): 11–28.

Farmer, J. Doyne and Arjun Makhijani. 2010. 'A US Nuclear Future? Counterpoint: Not Wanted, Not Needed'. *Nature* 467: 391–393, doi: 10.1038/467391a.

Farmer, J. Doyne and Spyros Skouras. 2013. 'An Ecological Perspective on the Future of Computer Trading'. *Quantitative Finance* 13 (3): 325–346, doi: 10.1080/14697688.2012.757636.

Farmer, J. Doyne and François Lafond. 2016. 'How Predictable Is Technological Progress?'. *Research Policy* 45 (3): 647–655, doi: 10.1016/j.respol.2015.11.001.

Farmer, J. Doyne, C. Hepburn, M. C. Ives, T. Hale, T. Wetzer, P. Mealy, R. Rafaty, S. Srivastav and R. Way. 2019. 'Sensitive Intervention Points in the Post-Carbon Transition'. *Science* 364 (6436): 132–134, doi: 10.1126/science.aaw7287.

Farmer, J. Doyne, Alissa M. Kleinnijenhuis, Paul Nahai-Williamson and Thom Wetzer. 2020. 'Foundations of System-Wide Financial Stress Testing with Heterogeneous Institutions'. Bank of England Working Paper No. 861.

Farmer, Roger E. A. October 4, 2016. 'Not Keen on More Chaos in the Future of Macroeconomics'. https://www.rogerfarmer.com/rogerfarmerblog/2016/10/4/nh0932exasra0c2a2amkvdmovcy9rz?rq=chaos

Fermi, Enrico, John Pasta, Stanislaw Ulam and Mary Tsingou. 1955. 'Studies of Nonlinear Problems'. Los Alamos National Laboratory Technical Report LA-1940, doi: 10.2172/4376203.

Fisher, Franklin. 1989. *Disequilibrium Foundations of Equilibrium Economics*. Cambridge: Cambridge University Press.

Forder, James. 2014. *Macroeconomics and the Phillips Curve Myth*. Oxford: Oxford University Press.

Fostel, A. and J. Geanakoplos. 2008. 'Leverage Cycles and the Anxious Economy'. *American Economic Review* 98 (4): 1211–44.

Foster, G. and P. Frijters. 2012. 'The Formation of Expectations: Competing Theories and New Evidence'. http://tinyurl.com/Foster-Frijters-2012

Frey, Carl Benedikt and Michael A. Osborne. 2013. 'The Future of Employment: How Susceptible are Jobs to Computerisation?' *Technological Forecasting and Social Change* 114 (C): 254–280, doi: 10.1016/j.techfore.2016.08.019.

Friedman, Milton. 1953. 'The Methodology of Positive Economics', in Milton Friedman (ed.), *Essays in Positive Economics*: 3–43. Chicago: University of Chicago Press.

Gabaix, X. and D. Laibson. 2008. 'The Seven Properties of Good Models', in Andrew Caplan and Andrew Schotter (eds), *The Foundations of Positive and Normative Economics: A Handbook*: 292–319. Oxford: Oxford University Press.

Gabaix, Xavier. 2011. 'The Granular Origins of Aggregate Fluctuations'. *Econometrica* 79 (3): 733–772.

Gabaix, Xavier. 2020. 'A Behavioral New Keynesian Model'. *American Economic Review* 110 (8): 2271–2327.

Galla, Tobias and J. Doyne Farmer. 2013. 'Complex Dynamics in Learning Complicated Games'. *Proceedings of the National Academy of Sciences* 110 (4): 1232–1236, doi: 10.1073/pnas.1109672110.

Galton, Francis. 1907. 'Vox Populi'. *Nature* 75: 450–451, doi: 10.1038/075450a0.

Geanakoplos, J. 1997. 'Promises, Promises . . .', in W.B. Arthur, S. Durlauf and D. Lane (eds), *The Economy as an Evolving Complex System, II*: 285–320. Reading: Addison-Wesley.

Geanakoplos, John. 2003. 'Liquidity, Default, and Crashes: Endogenous Contracts in General Equilibrium'. *Advances in Economics and Econometrics: Theory and Applications, Eighth World Conference, August 2000, Volume II, Econometric Society Monographs*: 170–205.

Geanakoplos, J. 2010. 'The Leverage Cycle', in D. Acemoglum, K. Rogoff and M. Woodford (eds.) *NBER Macroeconomics Annual 2009* (Vol. 24): 1–65. Chicago: University of Chicago Press.

Geanakoplos, John, Robert Axtell, J. Doyne Farmer, Peter Howitt, Benjamin Conlee, Jonathan Goldstein, Matthew Hendry, Nathan M. Palmer and Chun-Yi Yang. 2012. 'Getting at Systemic Risk via an Agent-Based Model of the Housing Market'. *American Economic Review* 102 (3): 53–58, doi: 10.1257/aer.102.3.53.

Gennaioli, Nicola and Andrei Shleifer. 2018. *A Crisis of Beliefs: Investor Psychology and Financial Fragility*. Princeton: Princeton University Press.

Gennotte, Gerard and Hayne Leland. 1990. 'Market Liquidity, Hedging, and Crashes'. *American Economic Review* 80 (5): 999–1021.

Ghashghaie, S., W. Breymann, J. Peinke, P. Talkner and Y. Dodge. 1996. 'Turbulent Cascades in Foreign Exchange Markets'. *Nature* 381: 767–770, doi: 10.1038/381767a0.

Ghazanfar, S. M., and A. Azim Islahi 1990. 'Economic Thought of an Arab Scholastic: Abu Hamid al-Ghazali.' *History of Political Economy* 22.2:381–403.

Gigerenzer, Gerd, Peter M. Todd and the ABC Research Group. 1999. *Simple Heuristics that Make Us Smart*. Oxford: Oxford University Press.

Gigerenzer, Gerd and Reinhard Selten (eds). 2001. *Bounded Rationality: The Adaptive Toolbox*. Cambridge, MA: MIT Press.

Gigerenzer, Gerd and Henry Brighton. 2009. 'Homo Heuristicus: Why Biased Minds Make Better Inferences'. *Topics in Cognitive Science* 1: 107–143, doi: 10.1111/j.1756-8765.2008.01006.x.

Gillemot, László, J. Doyne Farmer and Fabrizio Lillo. 2007. 'There's More to Volatility than Volume'. *Quantitative Finance* 6 (5): 371–384, doi: 10.1080/14697680600835688.

Glandon, Philip, Kenneth Kuttner, Sandeep Mazumder and Caleb Stroup. 2022. 'Macroeconomic Research Present and Past'. NBER Working Paper 29628.

Gleick, James. 1987. *Chaos: Making a New Science*. New York: Viking.

Gode, Dhananjay and Shyam Sunder. 1993. 'Allocative Efficiency of Markets with Zero-Intelligence Traders: Market as a Partial Substitute for Individual Rationality'. *Journal of Political Economy* 101 (1): 119–137, doi: 10.1086/261868.

Goldin, Claudia and Lawrence F. Katz. 2008. *The Race between Education and Technology*. Cambridge, MA: Harvard University Press.

Goodwin, Richard M. 1967, 'A Growth Cycle', in C. H. Feinstein (ed.), *Socialism, Capitalism and Economic Growth*. Cambridge: Cambridge University Press.

Gordon, Deborah M. 1999. *Ants at Work: How an Insect Society is Organized*. New York: The Free Press.

Grandmont, Jean-Michel. 1985. 'On Endogenous Competitive Business Cycles'. *Econometrica* 53 (5): 995–1045.

Granovetter, Mark S. 1973. 'The Strength of Weak Ties'. *American Journal of Sociology* 78 (6): 1360–1380, doi: 10.1086/225469.

Greenwood, Robin and David Scharfstein. 2013. 'The Growth of Finance'. *Journal of Economic Perspectives* 27 (2): 3–28, doi: 10.1257/jep.27.2.3.

Grossman, Sanford J. and Joseph E. Stiglitz. 1980. 'On the Impossibility of Informationally Efficient Markets'. *American Economic Review* 70 (3): 393–408.

Gualdi, Stanislao, Marco Tarzia, Francesco Zamponi and Jean-Philippe Bouchaud. 2015. 'Tipping Points in Macroeconomic Agent-Based Models'. *Journal of Economic Dynamics & Control* 50: 29–61.

Gurgone, Andrea. 2020. 'The Interaction of Borrower-Targeted Macroprudential Tools in the Irish Mortgage Market: A Baseline Multi-Agent Approach'. Working paper, https://papers.ssrn.com/sol3/papers.cfm?abstract_id=3708904.

Guvenin, Fatih, Fatih Karahan, Serdar Ozkan and Jae Song. 2021. 'What do Data on Millions of Workers Reveal About Lifecycle Earning Dynamics?'. *Econometrica* 89 (5): 2303–2339.

Helbing, D. and A. Kirman. 2013. 'Rethinking Economics Using Complexity Theory'. *Real-world Economics Review* 64.

Henrich, J., R. Boyd, S. Bowles, C. Camerer, E. Fehr and H. Gintis (eds). 2004. *Foundations of Human Sociality*. Oxford: Oxford University Press.

Hidalgo, Cesar A. and Ricardo Hausmann. 2009. 'The Building Blocks of Economic Complexity'. *Proceedings of the National Academy of Sciences* 106 (26): 10570–10575, doi:10.1073/pnas.0900943106.

Hill, Raymond R. and Andreas Tolk. 2017. 'A History of Military Computer Simulation', in Andreas Tolk et al. (eds), *Advances in Modeling and Simulation: Simulation Foundations, Methods, and Applications*. Cham: Springer, doi: 10.1007/978-3-319-64182-9_13.

Holland, John H. 2014. *Complexity: A Very Short Introduction*. Oxford: Oxford University Press.

Hommes, Cars. 2013. *Behavioral Rationality and Heterogeneous Expectations in Complex Economic Systems*. Cambridge: Cambridge University Press, doi:10.1017/CBO9781139094276.

Hommes, C., M. He, S. Poledna, M. Siqueira and Y. Zhang. 2022. *CANVAS: A Canadian Behavioral Agent-Based Model* (No. 2022-51). Bank of Canada.

IPCC. 2014. AR5 Sythesis Report. Climate Change 2014. https://www.ipcc.ch/report/ar5/syr/

IPCC. 2018. *Global Warming of 1.5 °C. An IPCC Special Report on the Impacts of Global Warming of 1.5 °C above Pre-industrial Levels and Related Global Greenhouse Gas Emission Pathways, in the Context of Strengthening the Global Response to the Threat of Climate Change, Sustainable Development, and Efforts to Eradicate Poverty*. V. Masson-Delmotte, P. Zhai, H. O. Pörtner, D. Roberts, J. Skea, P. R. Shukla, A. Pirani, W. Moufouma-Okia, C. Péan, R. Pidcock, S. Connors, J. B. R. Matthews, Y. Chen, X. Zhou, M. I. Gomis, E. Lonnoy, T. Maycock, M. Tignor, and T. Waterfield (eds). Cambridge and New York: Cambridge University Press, 616 pp., doi:10.1017/9781009157940.

Jackson, Matthew O., Brian Rogers and Yves Zenou. 2017. 'The Economic Consequences of Social Network Structure'. *Journal of Economic Literature* 55 (1): 49–95, doi: 10.1257/jel.20150694.

Jackson, T. 2016. *Prosperity Without Growth: Foundations for the Economy of Tomorrow*. Abingdon: Taylor & Francis.

Joulin, Armand, Augustin Lefevre, Daniel Grunberg and Jean-Philippe Bouchaud. 2008. 'Stock Price Jumps: News and Volume Play a Minor Role'. arXiv 0803.1769v1 [q-fin.ST].

Keynes, John Maynard. 1936. *The General Theory of Employment, Interest and Money*. London: Palgrave Macmillan.

Khandani, Amir E. and Andrew W. Lo. 2011. 'What Happened to the Quants in August 2007? Evidence from Factors and Transactions Data'. *Journal of Financial Markets* 14 (1): 1–46, doi: 10.1016/j.finmar.2010.07.005.

Kirman, A. P. 1992. 'Whom or What does the Representative Individual Represent?'. *Journal of Economic Perspectives* 6 (2): 117–136.

Kirman, A. 2010. 'The Economic Crisis is a Crisis for Economic Theory'. *CESifo Economic Studies* 56 (4): 498–535.

Knez, M. and V. Smith. 1987. 'Hypothetical Valuations and Preference Reversals in the Context of Asset Trading', in Alvin E. Roth (ed.), *Laboratory Experimentation in Economics: Six Points of View*: 131–154. Cambridge: Cambridge University Press.

Knight, Frank Hyneman. 1921. *Risk, Uncertainty, and Profit*. Boston: Houghton Mifflin.

Koh, Dongya, Raul Santaeulalia-Llopis and Yu Zheng. 2020. 'Labor Share Decline and Intellectual Property Products Capital'. *Econometrica* 88 (6): 2609–2628.

Koh, Heebyung and Christopher L. Magee. 2006. 'A Functional Approach for Studying Technological Progress: Application to Information Technology'. *Technological Forecasting and Social Change* 73: 1061–1083, doi: 10.1016/j.techfore.2006.06.001.

Ibid. 2008. 'A Functional Approach for Studying Technological Progress: Extension to Energy Technology', *Technological Forecasting and Social Change* 75 (6): 735–758, doi: 10.1016/j.techfore.2007.05.007.

Kolic, Blas, Juan Sabuco and J. Doyne Farmer. 2022. 'Estimating Initial Conditions for Dynamical Systems with Incomplete Information'. *Nonlinear Dynamics* 108: 3783–805, doi: 10.1007/s11071-022-07365-y.

Koopmans, T. C. 1965. 'On the Concept of Optimal Economic Growth', in L. Johnansen (ed.), *The Economic Approach to Development Planning*: 225–287. Chicago: Rand McNally.

Krugman, Paul. 1996. *The Self-Organizing Economy*. Oxford: Blackwell.

Ibid. April 29, 2015. 'The Case for Cuts was a Lie. Why Does Britain Still Believe It? The Austerity Delusion'. *The Guardian*. https://www.theguardian.com/business/ng-interactive/2015/apr/29/the-austerity-delusion

Ibid. June 29, 2015. 'Greece over the Brink'. *The New York Times*. https://www.nytimes.com/2015/06/29/opinion/paul-krugman-greece-over-the-brink.html

Ladyman, James and Karoline Wiesner. 2020. *What is a Complex System?* New Haven: Yale University Press.

Lafond, François, Aimee Gotway Bailey, Jan David Bakker, Dylan Rebois, Rubina Zadourian, Patrick McSharry and J. Doyne Farmer. 2018. 'How Well Do Experience Curves Predict Technological Progress? A Method for Making Distributional Forecasts'. *Technological Forecasting and Social Change* 128: 104–117, doi: 10.1016/j.techfore.2017.11.001.

Lafond, François, Diana Seave Greenwald and J. Doyne Farmer. 2020. 'Can Stimulating Demand Drive Costs Down? World War II as a Natural Experiment'. INET Oxford Working Paper No. 2020-02, https://www.inet.ox.ac.uk/publications/no-2020-02-can-stimulating-demand-drive-costs-down-world-war-ii-as-a-natural-experiment/

Lamperti, F., G. Dosi, M. Napoleatano, A. Roventini and A. Sapio. 2018. 'Coupled Climate and Economic Dynamics in an Agent-based Integrated Assessment Model'. *Ecological Economics* 150: 315–339.

Lazo, J. K., R. E. Morss and J. L. Demuth. 2009. '300 Billion Served: Sources, Perceptions, Uses and Values of Weather Forecasts', *Bulletin of the American Meteorological Society* 90: 785–798, doi: 10.1175/2008BAMS2604.1.

Legendre, Adrien-Marie. 1806. *Nouvelles méthodes pour la détermination des orbites des comètes*. Paris: Firmin Didot.

Leontief, Wassily W. 1951. 'Input-Output Economics'. *Scientific American* 185 (4): 15–21.

Levin, S. A. and A. W. Lo. 2021. 'Introduction to PNAS Special Issue on Evolutionary Models of Financial Markets'. *Proceedings of the National Academy of Sciences* 118 (26): e2104800118.

Li, T. J. and J. A. Yorke. 1975. 'Period Three Implies Chaos'. *American Mathematical Monthly* 82: 985–992.

Lo, A.W. 2004. 'The Adaptive Markets Hypothesis: Market Efficiency from an Evolutionary Perspective'. *Journal of Portfolio Management*. 2004.

Lo, A.W. and R. Zhang. 2018. 'Biological economics', in *Biological Economics*. Edward Elgar Publishing.

Lo, Andrew W. 2019. *Adaptive Markets*. Princeton: Princeton University Press. https://press.princeton.edu/books/paperback/9780691191362/adaptive-markets

Lorenz, Edward Norton. 1963. 'Deterministic Nonperiodic Flow'. *Journal of the Atmospheric Sciences* 20 (2): 130–141, doi: 10.1177/0309133315623099.

Lotka, Alfred J. 1925. *Elements of Physical Biology*. Baltimore: Williams and Wilkins.

Mantegna, R. N. and H. E. Stanley. 1999. *Introduction to Econophysics: Correlations and Complexity in Finance*. Cambridge: Cambridge University Press.

Markowitz, Harry. 1952. 'Portfolio Selection'. *Journal of Finance* 7 (1): 77–91, doi: 10.2307/2975974.

Marshall, Alfred. 1920. *Principles of Economics: An Introductory Volume*, eighth ed. London: Macmillan and Co.

Marx, Karl. 1867. *Capital Volume I*, reissue ed: 1990, London: Penguin Classics.

McNerney, James, J. Doyne Farmer, Sidney Redner and Jessika E. Trancik. 2011. 'Role of Design Complexity in Technology Improvement'. *Proceedings of the National Academy of Sciences* 108 (22): 9008–9013, doi: 10.1073/pnas.1017298108.

McNerney, J., C. Savoie, F. Caravelli, V. M. Carvalho and J. D. Farmer. 2022. 'How Production Networks Amplify Economic Growth'. *Proceedings of the National Academy of Sciences* 119 (1): e2106031118, doi: 10.1073/pnas.2106031118

Mealy, Penny, R. Maria del Rio-Chanona and J. Doyne Farmer. 2018. 'What You Do at Work Matters: New Lenses on Labour'. Working paper. https://papers.ssrn.com/sol3/papers.cfm?abstract_id=3143064

Meng, Jing, Rupert Way, Elena Verdolini and Laura Diaz Anadon. 2021. 'Comparing Expert Elicitation and Model-Based Probabilistic Technology Cost Forecasts for the Energy Transition'. *Proceedings of the National Academy of Sciences* 118 (27): e1917165118, doi: 10.1073/pnas.1917165118.

Mérő, Bence, András Borsos, Zsuzsanna Hosszú-Zsolt and Oláh-Nikolett Vágó. 2022. 'A High Resolution Agent-based Model of the Hungarian Housing Market'. MNB Working papers 7.

Miller, John H. and Scott E. Page. 2007. *Complex Adaptive Systems: An Introduction to Computational Models of Social Life.* Princeton: Princeton University Press.

Mirowski, Philip. 1989. *More Heat than Light: Economics as Social Physics, Physics as Nature's Economics*. Cambridge: Cambridge University Press.

Mitchell, Melanie. 2009. *Complexity: A Guided Tour*. Oxford: Oxford University Press.

Moore, Gordon E. 1965. 'Cramming More Components into Integrated Circuits'. *Electronics* 38 (8): 114.

Moses, M.E., S. Hofmeyr, J. L. Cannon, A. Andrews, R. Gridley, M. Hinga, K. Leyba, A. Pribisova,V. Surjadidjaja, H. Tasnim and S. Forrest. 2021. 'Spatially Distributed Infection Increases Viral Load in a Computational Model of SARS-CoV-2 Lung Infection'. *PLoS Computational Biology* 17 (12): e1009735.

Mungo, Luca, François Lafond, Pablo Astudillo-Estévez and J. Doyne Farmer. 2022. 'Reconstructing Production Networks Using Machine Learning'. INET Oxford Working Paper No. 2022-02.

Muth, John F. 1961. 'Rational Expectations and the Theory of Price Movements'. *Econometrica* 29 (3): 315–335, doi: 10.2307/1909635.

Muzy, J., J. Delour and E. Bacry. 2000. 'Modelling Fluctuations of Financial Time Series: From Cascade Process to Stochastic Volatility Model'. *European Physics Journal B* 17: 537–548. https://doi.org/10.1007/s100510070131

Nagy, Béla, J. Doyne Farmer, Quan M. Bui and Jessika E. Trancik. 2013. 'Statistical Basis for Predicting Technological Progress'. *PLoS ONE* 8 (2): e52669, doi: 10.1371/journal.pone.0052669.

Nash, John. 1951. 'Non-Cooperative Games'. *The Annals of Mathematics*, Second Series 54 (2): 286–295.

National Research Council. 1979. *Carbon Dioxide and Climate: A Scientific Assessment*. Washington, DC: The National Academies Press, doi: 10.17226/12181.

Nelson, Richard R. and Sidney G. Winter. 1982. *An Evolutionary Theory of Economic Change*. Cambridge, MA: The Belknap Press of Harvard University Press.

Newman, Mark. 2018. *Networks*, second edition. Oxford: Oxford University Press.

Newman, M. E. J. 2018. 'Network Structure from Rich but Noisy Data'. *Nature Physics* 14: 542–45.

Nordhaus, William D. 1991. 'To Slow or Not to Slow: The Economics of the Greenhouse Effect'. *Economic Journal* 101 (407): 920–937, doi: 10.2307/2233864.

Ibid. 1992. 'An Optimal Transition Path for Controlling Greenhouse Gases.' *Science* 258 (5086): 1315–1319, doi: 10.1126/science.258.5086.1315.

Ibid. 2017. 'Revisiting the Social Cost of Carbon,' *PNAS* 114 (7): 1518–1523, doi: 10.1073/pnas.1609244114.

O*NET OnLine, National Center for O*NET Development, www.onetonline.org/. (Accessed 13 April 2022.)

Opper, Manfred and Sigurd Diederich. 1992. 'Phase Transition and 1/f Noise in a Game Dynamical Model'. *Physical Review Letters* 69 (10): 1616–1619, doi: 10.1103/PhysRevLett.69.1616.

Packard, N. H., J. P. Crutchfield, J. D. Farmer and R. S. Shaw. 1980. 'Geometry from a Time Series'. *Physical Review Letters*, 45 (712): 712–716, doi: 10.1103/PhysRevLett.45.712.

Page, Lawrence, Sergey Brin, Rajeev Motwani and Terry Winograd. 1999. 'The PageRank Citation Ranking: Bringing Order to the Web'. Technical Report. Stanford InfoLab.

Page, Scott E. 2007. *The Difference*. Princeton: Princeton University Press.

Ibid. 2018. *The Model Thinker: What You Need to Know to Make Data Work for You*. New York: Basic Books.

Palmer, R. G., W. B. Arthur, John H. Holland, Blake LeBaron and Paul Tayler. 1994. 'Artificial Economic Life: A Simple Model of a Stockmarket'. *Physica D* 75 (1): 264–274, doi: 10.1016/0167-2789(94)90287-9.

Pangallo, Marco, James B. T. Sanders, Tobias Galla and J. D. Farmer. 2022. 'Towards a Taxonomy of Learning Dynamics in 2 x 2 Games'. *Games and Economic Behavior* 132: 1–21.

Pangallo, Marco, Torsten Heinrich and J. D. Farmer. 2019. 'Best Reply Structure and Equilibrium Convergence in Generic Games'. *Science Advances* 5 (2), doi: 10.1126/sciadv.aat1328.

Pangallo, M., A. Aleta, R. Chanona, A. Pichler, D. Martín-Corral, M. Chinazzi, F. Lafond, M. Ajelli, E. Moro, Y. Moreno and A. Vespignani. 2023. 'The Unequal Effects of the Health-Economy Tradeoff During the COVID-19 Pandemic'. *Nature Human Behaviour*, pp.11–12.

Philippon, Thomas and Arielle Reshef. 2009. 'Wages and Human Capital in the U.S. Financial Industry: 1909–2006'. NBER working paper 14644, doi: 10.3386/w14644.

Pichler, Anton, Marco Pangallo, R. Maria del Rio-Chanona, François Lafond and J. Doyne Farmer. 2020. 'Production Networks and Epidemic Spreading: How to Restart the UK Economy?' arXiv: 2005.10585 [econ.GN], doi: 10.48550/arXiv.2005.10585.

Pichler, Anton, Marco Pangallo, R. Maria del Rio-Chanona, François Lafond and J. Doyne Farmer. June 7, 2020. 'Production Networks and Epidemic Spreading: Re-Opening the UK Economy'. VoxEU.org. https://voxeu.org/article/production-networks-and-epidemic-spreading-re-opening-uk-economy

Pichler, A., M. Pangallo, R. M. del Rio-Chanona, F. Lafond and J. D. Farmer. 2022. 'Forecasting the Propagation of Pandemic Shocks with a Dynamic Input-Output Model'. *Journal of Economic Dynamics and Control* Volume 144:104527. https://doi.org/10.1016/j.jedc.2022.104527

Pichler, Anton, François Lafond and J. Doyne Farmer. 2020. 'Technological Interdependencies Predict Innovation Dynamics'. arXiv: 2003.00580 [physics.soc-ph], doi: 10.48550/arXiv.2003.00580.

Piketty, Thomas. 2015. *The Economics of Inequality*. Cambridge, MA: Harvard University Press.

Piketty, Thomas, Emmanuel Saez and Gabriel Zucman. 2018. 'Distributional National Accounts: Methods and Estimates from the United States'. *Quarterly Journal of Economics* 133 (2): 553-609.

Platt, Donovon. 2022. 'Bayesian Estimation of Economic Simulation Models Using Neural Networks'. *Computational Economics* 59: 599–650, doi: 10.1007/s10614-021-10095-9.

Plonsky, Ori, Reut Apel, Eyal Ert, Moshe Tennenholtz, David Bourgin, Joshua C. Peterson, Daniel Reichman, Thomas L. Griffiths, Stuart J. Russell, Evan C. Carter, James F. Cavanagh and Ido Erev. 2019. 'Predicting Human Decisions with Behavioral Theories and Machine Learning'. arXiv: 1904.06866 [cs.AI], doi: 10.48550/arXiv.1904.06866.

Poincaré, Henri. 1914. *Science and Method*, Chapter 4. T. Nelson & Sons.

Polanyi, Karl. *The Great Transformation: the Political and Economic Origins of our Time*. Foreword by Joseph E. Stiglitz; with a new introd. by Fred Block.—2nd Beacon Paperback ed. p. cm. Originally published: New York: Farrar & Rinehart, 1944 and reprinted in 1957 by Beacon in Boston.

Poledna, Sebastian, Stefan Thurner, J. Doyne Farmer and John Geanakoplos. 2014. 'Leverage-Induced Systemic Risk under Basel II and other Credit Risk Policies'. *Journal of Banking & Finance* 42: 199–212, doi: 10.1016/j.jbankfin.2014.01.038.

Poledna, Sebastian, Michael Gregor Miess, Cars Hommes and Katrin Rabitsch. 2022. 'Economic Forecasting with an Agent-Based Model'. *European Economic Review*, 104306, doi: 10.1016/j.euroecorev.2022.104306.

Ramsey, Frank P. 1928. 'A Mathematical Theory of Saving'. *Economic Journal* 38 (152): 543–559, doi:10.2307/2224098.

Reinhart, Carmen M. and Kenneth S. Rogoff. 2009. *This Time is Different: Eight Centuries of Financial Folly*. Princeton: Princeton University Press.

Richardson, Lewis Fry. 1922. *Weather Prediction by Numerical Process*. London: Cambridge University Press.

Rissanen, Jorma. 2007. *Information and Complexity in Statistical Modeling*. Cham: Springer.

Romer, Paul M. 2016. 'The Trouble with Macroeconomics'. Working paper, https://paulromer.net/trouble-with-macroeconomics-update/WP-Trouble.pdf.

Ruelle, David and Floris Takens. 1971. 'On the Nature of Turbulence'. *Communications in Mathematical Physics* 20: 167–192, doi: 10.1007/BF01646553.

Sachs, J. D., G. Schmidt-Traub, M. Mazzucato, D. Messner, N. Nakicenovic and J. Rockström. 2019. 'Six Transformations to Achieve the Sustainable Development Goals'. *Nature Sustainability* 2 (9): 805–814.

Samuelson, Paul A. 1974. 'A Biological Least-Action Principle for the Ecological Model of Lotka–Volterra' *Proceedings of the National Academy of Sciences USA* 71 (8): 3041–3044, doi: 10.1073/pnas.71.8.3041.

Sanders, James B. T., J. Doyne Farmer and Tobias Galla. 2018. 'The Prevalence of Chaotic Dynamics in Games with Many Players'. *Scientific Reports* 8: 4902, doi: 10.1038/s41598-018-22013-5.

Sahal, Devendra. 1982. *The Transfer and Utilization of Technical Knowledge*. Lexington: Lexington Books.

Sargent, T. J., 1993. 'Bounded Rationality in Macroeconomics: The Arne Ryde Memorial Lectures'. *OUP Catalogue*.

Sato, Yuzuru, Eizo Akiyama and J. Doyne Farmer. 2002. 'Chaos in Learning a Simple Two-Person Game'. *Proceedings of the National Academy of Sciences* 99 (7): 4748–4751, doi: 10.1073/pnas.032086299.

Savoie, Charles. 2017. 'Input-Output Analysis and Growth Theory'. PhD thesis, University of Oxford, Lincoln College.

Scheinkman, José Alexandre. 1976. 'On Optimal Steady States of *n*-sector Growth Models when Utility is Discounted'. *Journal of Economic Theory*, 12 (1): 11–30.

Scholl, Maarten P., Anisoara Calinescu and J. Doyne Farmer. 2021. 'How Market Ecology Explains Market Malfunction'. *Proceedings of the National Academy of Sciences* 118 (26): e2015574118, doi: 10.1073/pnas.2015574118.

Schrödinger, Erwin. 1948. *What is Life? The Physical Aspect of the Living Cell*. Cambridge: Cambridge University Press.

Schumpeter, Joseph. 1994 [1942]. *Capitalism, Socialism, and Democracy*. London: Routledge.

Shannon, C.E. 1948. 'A Mathematical Theory of Communication'. *Bell System Technical Journal* 27: 379–423, 623–656, doi: 10.1002/j.1538-7305.1948.tb01338.x.

Shiller, Robert J. 1981. 'Do Stock Prices Move Too Much to be Justified by Subsequent Changes in Dividends?' *American Economic Review* 71 (3): 421–436.

Simon, Herbert A. 1955. 'A Behavioral Model of Rational Choice'. *Quarterly Journal of Economics* 69 (1): 99–118, doi: 10.2307/1884852.

Ibid. 1957. *Models of Man: Social and Rational – Mathematical Essays on Rational Human Behavior in a Social Setting*. New York: John Wiley & Sons.

Sims, Chris. 1980. 'Macroeconomics and Reality'. *Econometrica* 48 (1): 4.

Smets, Frank and Rafael Wouters. 2007. 'Shocks and Frictions in US Business Cycles: A Bayesian DSGE Approach'. *American Economic Review* 97 (3): 586–606, doi: 10.1257/aer.97.3.586.

Smith, Adam. 1776. *The Wealth of Nations*. London: W. Strahan & T. Cadell.

Smith, Eric, J. Doyne Farmer, László Gillemot and S. Krishnamurthy. 2002. 'Statistical Theory of the Continuous Double Auction'. *Quantitative Finance* 3 (6): 481–514, doi: 10.1088/1469-7688/3/6/307.

Smith, V.L. 1962. 'An Experimental Study of Competitive Market Behavior'. *Journal of Political Economy* 70 (2): 111–137.

Smith, Vernon. 2008. *Rationality in Economics: Constructivist and Ecological Forms*. New York: Cambridge University Press.

Solow, Robert M. 1956. 'A Contribution to the Theory of Economic Growth'. *Quarterly Journal of Economics* 70 (1): 65–94, doi: 10.2307/1884513.

Stanley, H. Eugene, V. Afanasyev, L. A. N. Amaral, S. V. Buldyrev, A. L. Goldberger, S. Havlin, H. Leschhorn, P. Maass, R. N. Mantegna, C. K. Peng, P. A. Prince, M. A. Salinger, M. H. R. Stanley and G. M. Viswanathan. 1996. 'Anomalous Fluctuations in the Dynamics of Complex Systems: From DNA and Physiology to Econophysics'. *Physica A* 224 (1–2): 302–321, doi: 10.1016/0378-4371(95)00409-2.

Stein, P. R. and Stanislaw Ulam. 1964. *Non-Linear Transformation Studies on Electronic Computers*. Warsaw: Panstwowe Wydawnictwo Naukowe.

Stern, Nicholas H. 2006. *The Economics of Climate Change: The Stern Review*. Cambridge: Cambridge University Press.

Stern, Nicholas, Joseph E. Stiglitz and Charlotte Taylor. 2021. 'The Economics of Immense Risk, Urgent Action, and Radical Change: Towards New Approaches to the Economics of Climate Change'. NBER Working Paper 28472, https://www.nber.org/paper65s/w28472.

Stiglitz, Joseph. November 19, 2006. 'A Cool Calculus of Global Warming'. *Project Syndicate*. https://www.project-syndicate.org/commentary/a-cool-calculus-of-global-warming-2006-11

Strober, Elizabeth A. 1998. 'Interview with Barbara Bergmann'. *Feminist Economics* 4 (3): 5–6.

Summers, L. H. 1986. 'Does the Stock Market Rationally Reflect Fundamental Values?' *Journal of Finance* 41 (3): 591–601.

Ibid. 2000. 'International Financial Crises: Causes, Prevention, and Cures'. *American Economic Review*, 90 (2): 1–16.

Tarne, Ruben, Dirk Bezemer and Thomas Theobald. 2021. 'The Effect of Borrower-Specific Loan-to-Value Policies on Household Debt, Wealth Inequality and Consumption Volatility: An Agent-Based Analysis'. FMM Working paper No. 70. https://papers.ssrn.com/sol3/papers.cfm?abstract_id=3915516.

Tesfatsion, Leigh. 2006. 'Agent-Based Computational Economics: A Constructive Approach to Economic Theory' in *Handbook of Computational Economics* (Vol. 2): 831–880.

Theiler, James, Stephen Eubank, André Longtin, Bryan Galdrikian and J. Doyne Farmer. 1992. 'Testing for Nonlinearity in Time Series: The Method of Surrogate Data'. *Physica D* 58 (1–4): 77–94, doi: 10.1016/0167-2789(92)90102-S.

Thorp, Edward O. and Sheen T. Kassouf. 1967. *Beat the Market: A Scientific Stock Market System*. New York: Random House.

Thurner, Stefan, J. Doyne Farmer and John Geanakoplos. 2012. 'Leverage Causes Fat Tails and Clustered Volatility'. *Quantitative Finance* 12 (5): 695–707, doi: 10.1080/14697688.2012.674301.

Thurner, Stefan, Rudolf Hanel and Peter Klimek. 2018. *Introduction to the Theory of Complex Systems*. Oxford: Oxford University Press.

Trichet, Jean-Claude. 2010. Opening Address, ECB Central Banking Conference Frankfurt, November 18, 2010 https://www.ecb.europa.eu/press/key/date/2010/html/sp101118.en.html.

Turchin, P. 2023. *End Times: Elites, Counter-Elites, and the Path of Political Disintegration*. New York: Penguin Press.

Veblen, Thorstein. 1899. *The Theory of the Leisure Class*. New York: The Macmillan Company.

Volterra, Vito. 1926. 'Variazioni e fluttuazioni del numero d'individui in specie animali conviventi'. *Mem. Accademia dei Lincei Roma* 2: 31–113.

von Neumann, John. 1966. *Theory of Self-Reproducing Automata*, edited and completed by Arthur W. Burks. Urbana: University of Illinois Press.

Way, Rupert, Matthew C. Ives, Penny Mealy and J. Doyne Farmer. 2022. 'Empirically Grounded Technology Forecasts and the Energy Transition'. *Joule* 6 (9): 2057–2082, doi: 10.1016/j.joule.2022.08.009.

Wicksell, Knut. 1918. 'Review: Ett bidrag till krisernas teori'. *Ekonomisk Tidskrift* 20 (2): 66–75, doi: 10.2307/3437495.

Wiener, Norbert. 1948. *Cybernetics: Or Control and Communication in the Animal and the Machine*. Cambridge, MA: MIT Press.

Wiersema, Garbrand, Alissa M. Kleinnijenhuis, Thom Wetzer and J. Doyne Farmer. 2023. 'Scenario-Free Analysis of Financial Stability with Interacting Contagion Channels'. *Journal of Banking & Finance* 146: 106684.

Wright, T. P. 1936. 'Factors Affecting the Cost of Airplanes'. *Journal of the Aeronautical Sciences* 3: 4, doi: 10.2514/8.155.

Zhou, Ping. September 1, 2021. 'Australia's Massive Feral Rabbit Problem'., *ThoughtCo.* thoughtco.com/feral-rabbits-in-australia-1434350.

Zucman, G. and E. Saez. 2019. *The Triumph of Injustice: How the Rich Dodge Taxes and How to Make them Pay*. New York: W. W. Norton & Company.

Notes

Prologue

1 Phelps, Edmund S. (2018). 'Equilibrium: An Expectational Concept.' *The New Palgrave Dictionary of Economics*. Palgrave Macmillan, London. https://doi.org/10.1057/978-1-349-95189-5_334

2 Arnold et al. (2006). https://www.cbo.gov/sites/default/files/109th-congress-2005-2006/reports/12-08-birdflu.pdf

3 https://www.onetonline.org

4 https://www.governo.it/sites/new.governo.it/files/dpcm_20200322.pdf

5 del Rio-Chanona et al. (2020).

6 O*NET prompted us to target the US economy, but the disadvantage was that US policies for dealing with COVID differed from state to state, and the lockdowns were far from simultaneous. So, while the first paper (predicting only the first-order shocks) targeted the US, the second paper, which predicted how the shocks would be amplified as they moved through the economy, focused on the UK.

7 To determine which inputs are critical, we convinced IHS Markit to survey their industry analysts. One or more of their analysts flagged the essential inputs for each of the fifty-two different industries. This gave us a much higher level of realism than other models, which was essential to the success of our predictions.

8 Pichler et al. (2020, 2022). We posted our paper publicly on May 21, 2020. The version we posted then had essentially the same results as the May 8 version, but the 2022 version does a post-mortem, examining our forecasts for each industry and a few time horizons. The conclusion of our post-mortem was that, though we did have a bit of luck, we mostly succeeded due to skill.

9 Alleman et al. (2023).

10 Farmer (2005).

Introduction

1 Strober, Elizabeth A. (1998) Interview with Barbara Bergmann. 4:3, 5–6, DOI: 10.1080/135457098338284
2 If you have any doubt that the economy is a major driver of politics and social change, I highly recommend Peter Turchin's book *End Times* (2023).
3 See https://www.federalreserve.gov/econres/us-models-about.htm
4 From a presentation given by Simon Potter, then Director of Economic Research at the New York Fed, at a 2010 conference I co-organized with Rob Axtell, titled 'Agent-Based Models of the Financial Crisis'.
5 See, for example, Trichet (2010).
6 See Piketty et al. (2018), Koh et al. (2020), Guvenin et al. (2021) and Buda et al. (2022).
7 Miller and Page (2007), Mitchell (2009), Holland (2014), Thurner, Hanel and Klimek (2018) and Ladyman and Wiesner (2020).
8 Arthur (1999, 2021). Alan Kirman is another important pioneer of complexity economics, see Kirman (1992), Kirman (2010) and Helbing and Kirman (2013). An early agent-based modeler whose work I will not be able to discuss here is Leigh Tesfatsion (2006).
9 See Miller and Page (2007).
10 For a review, see Axtell and Farmer (2022).
11 Gigerenzer and Brighton (2009).
12 Page (2007)
13 Del Negro et al. (2017).
14 Several economists, like Robert Shiller, Raghuram Rajan, Nouriel Roubini and others (including Simon Potter and the economists at the New York Fed discussed at the beginning of the introduction), presciently warned of a crisis in the years leading up to 2007. But their warnings were based on good economic intuition rather than theoretical models. Their warnings were also confined to the possibility of financial crisis – almost everyone failed to anticipate the broader consequences for the economy as a whole.

15 See Brunnermeier and Sannikov (2014) and Del Negro et al. (2017).

16 A model that comes close to being comprehensive, taking many different economic factors into account at once, is FRB/US, but it is an econometric model – while it has economic constraints built into it, it is not derived from a theory based on utility maximization.

17 Friedman (1953).

18 Smith (1776).

19 Nordhaus (1992, 2017).

20 In the US there are only two complexity economists who sit in economics departments. Europe is more supportive, but departments that allow complexity economics are still rare.

Part I: What is Complexity Economics?

1 At the time my Stanford tuition was $2,400 a year, which meant that I was paying my professors about $3 an hour for their lectures. This was roughly the same wage I got working summers doing manual labor, which made me feel justified in asking questions – I wanted my money's worth.

2 Fisher (1989).

3 In the process of *tatonnement* an auctioneer picks a price and queries the audience for their offers to buy or sell at that price. If the two do not match, he chooses another price and repeats the process. He continues this until supply equals demand, at which point transactions are made.

4 Anderson, Arrow and Pines (1988).

5 Farmer and Geanakoplos (2009).

1: What is a Complex System?

1 Anderson, P. W. 2000. 'Emergence, Reductionism and the Seamless Web: When and Why Is Science Right.' *Current Science*, vol. 78, no. 6, 2000, pp. 673–76. JSTOR, http://www.jstor.org/stable/24103876.

2 See Mitchell (2009), Holland (2014) and Thurner et al. (2018).

3 See Wiener (1948), Shannon (1948), von Neumann (1966) and Schrödinger (1948).

4 Farmer et al. (1984), Farmer et al. (1986c).

5 There is no precise, objective mathematical definition of an emergent property; at this stage, the classification remains subjective – a situation reminiscent of Justice Potter Stewart's famous statement about pornography in *Jacobellis v. Ohio* (1964): 'I know it when I see it.'

6 Gordon (1999).

7 Krugman (2015a, 2015b).

8 Harry Truman famously said, 'Give me a one-handed economist. All my economists say "on one hand", then "but on the other".'

9 Gabaix and Laibson (2008).

10 Romer (2016).

11 Bass (1985).

12 For an account about who Tom Ingerson is and how he inspired me to become a scientist, see Farmer (2004).

13 This also depends on the wheel being in good working order. If one or more of the frets separating the cups on the rotor sticks up too much, the number in front of it will come up more often than it should. This has been successfully used to beat the house. But if the casino takes good care of their wheels this doesn't happen.

14 There is also rolling friction, caused by the contact of the ball against the wheel, but our experiments made it clear that this is small in comparison to wind resistance. Rolling friction is simpler to handle because it is constant, in contrast to wind resistance, which depends on velocity. The slowing down of the spinning wheel in the center can be predicted using the assumption of rolling friction, but the ball cannot be predicted very well this way.

15 In part because of our demonstration that roulette could be beaten, the state of Nevada subsequently passed a law against 'using a computer to predict the outcome of a game'.

16 As we were completing construction of the first wearable digital computer, our friend Jim Crutchfield came home one evening and said that he had just been to a meeting of the HomeBrew Computer Club,

where two guys were working on a computer that they said they were going to sell to housewives. We couldn't figure out why housewives would want computers, and decided that we would make more money by beating roulette. The two guys were Steve Jobs and Steve Wozniak.

17 Lorenz (1963).

18 The term 'chaos' for sensitive dependence on initial conditions was first used by Li and Yorke in a 1975 paper called 'Period Three Implies Chaos'. However, it didn't really catch on until the 1980s, in particular with the publication of James Gleick's popular book, *Chaos* (1987).

19 Poincaré (1914).

20 Packard et al. (1980).

21 Wicksell, Knut. 1918. 'Review of Karl Petander, "Goda och da ~liga tider".' *Ekonomisk Tidskrift* 11: 66–75.

22 See, for example, Benhabib and Nishimura (1979), Benhabib and Day (1981), Day (1982), Grandmont (1985) and Diamond and Fudenberg (1989).

23 Beaudry, Galizia and Portier (2015, 2016).

24 Day (1981) and Boldrin and Montrucchio (1986).

25 Brock et al. (1991).

26 See R. Farmer (2016) for an account of how chaos and the events narrated here are viewed by the economics community.

27 In the late 1980s, several papers claimed that EEGs displayed low dimensional chaos. In Theiler et al. (1992) we extended the techniques developed by Brock et al. to allow for correlations, and produced a similar result showing that in the case of the EEG there was no evidence to support this hypothesis.

28 Ruelle and Takens (1971).

29 To give a bit more detail about the Ruelle and Takens' theorem: As discussed in Chapter 1, there are only two types of endogenous motion: regular motion and chaos. Regular motion corresponds to a limit cycle or a combination of limit cycles with frequencies whose ratios are irrational numbers. By combining two such cycles, you get a motion that looks like a doughnut, which mathematicians call a 2-torus. Similarly, if you combine three such cycles, you get a

higher-dimensional generalization of a doughnut, called a 3-torus. And so on. Ruelle and Takens proved that anything higher than a 2-torus is very unlikely to occur. That is, if you make up dynamical systems at random, their attractors can be fixed points or limit cycles or 2-tori, but they will almost never be a 3-torus or higher. Instead, chaotic attractors become likely. They used this theorem to argue that chaotic attractors must underlie turbulence.

2: The Economy is a Complex System

1 Krugman, Paul. 1996. *The Self Organizing Economy*. Wiley Blackwell.
2 Krugman (1996).
3 When Malthus observed that the population grows exponentially, he failed to realize that the real limit is the carrying capacity of the earth to support the human population, which has dramatically increased through time. Malthus's essay was a comment on Marquis de Condorcet's essay, *Sketch for a Historical Picture of the Progress of the Human Mind*. Unlike Malthus, Condorcet saw clearly that technological progress was increasing the carrying capacity for the human population and correctly predicted that it would continue to grow rapidly as a result.
4 See, for example, Lo and Zhang (2018).
5 Marshall (1920), preface to the 8th edition.
6 Schumpeter (1942).
7 Nelson and Winter (1982), p.4.
8 Smith (1776). See book 1, chapter 3. As this book is going to press I learned that there is a much older, very articulate statement of how the economy operates through the interaction of specialists by Ilya Ulum al Din al-Ghazali, who lived from 1058–1111. See Ghazanfar and Illahi (1990).
9 For organisms, the energy budget plays a role that is in some ways analogous to the balance sheet; organisms in an ecosystem interact via their energy flows, for example by eating or being eaten.

3: Understanding the Metabolism of Civilization

1 Karl Marx, *Capital*, Volume 1, reissue ed. 1990, London: Penguin Classics.
2 Self-organization is a particular kind of emergence in which disorganized configurations of matter become highly organized. The origin of life is a good example.
3 Granovetter (1973).
4 See, for example, Newman (2018a, 2018b).
5 See, for example, Jackson, Rogers and Zenou (2017).
6 Page et al. (1999).
7 Battiston et al. (2012).
8 Farmer et al. (1986a).
9 Farmer et al. (1986b).
10 Bagley and Farmer (1991).
11 Leontief (1951).
12 Thinking in terms of money flows has an important drawback: It fails to give any insight into sustainability. The flow of material and energy is not circular – the economy gathers raw materials and produces waste, so material flows into and out of the economy. But the flow of money is circular by definition. Mining companies, for example, extract raw materials and sell them, but their money flows look much like those of any other company: The money flowing in from sales equals the money flowing out for expenses and profits. Unfortunately, almost all of the measurements of the production network are money flows rather than flows of material and energy.
13 Clark (1956), foreword by Wassily Leontief.
14 Stanley et al. (1996).
15 Gabaix (2011).
16 McNerney et al. (2022).
17 Households consume the goods that the economy produces, and they are analogous to the organism that a metabolism feeds. Households also provide labor to the economy, much as an organism gathers food for the metabolism to process.

18 The trophic level is also equivalent to the average number of times a dollar (or renminbi) spent by an industry changes hands until it ends up in the pocket of a worker.

19 If the government injects extra demand, for example by subsidizing household food purchases, the output multiplier measures how much this increases overall output in the economy. We are using the concept differently, but this is equivalent to what I am calling the trophic level here.

20 Baumol (2012).

21 Our predictions are about as accurate as other, earlier predictions based on factors like GDP, education and labor share. However, because none of these predictions are very accurate, and because our predictions are right when others are wrong and vice versa, if you average our predictions together with existing methods of prediction, they result in an overall improvement in predictive power. (This follows a well-known principle discovered by Francis Galton in the twentieth century, that is now sometimes called 'the wisdom of crowds' (Galton, 1907) – but in this case the 'crowd' corresponds to different models.) See Page (2018).

22 Savoie (2017).

23 The trophic levels of the economy are a result of all the decisions made by producers and consumers through time, so in a sense they depend on human decisions. But to make our predictions we didn't need to worry about how the trophic levels were created – we just took them as given.

24 We derived the model in McNerney et al. (2022) from a basic statement in accounting: The amount that companies pay for their inputs is equal to their revenue from selling their outputs minus their profits. We assumed the profits are small enough to be negligible in comparison to the revenue, which is not a bad approximation relative to the other factors influencing technological innovation. Our derivation allows any production function and makes no assumptions about utility maximization.

25 The standard way to derive the basic model requires the assumption that firms maximize the utility of their profits and that the relationship between the inputs and outputs is given by a Cobb-Douglas

production function. The Cobb-Douglas assumption is reasonable for highly aggregate circumstances involving the tradeoff between capital and labor, but it is not a plausible description of production functions needed at the level of industries or products.

26 Beinhocker (2007). Bookstaber (2019) is another nice book advocating agent-based modeling in economics, and see Bouchaud (2008) for an articulate manifesto for complexity economics.

27 Mealy, del Rio-Chanona and Farmer (2018).

28 del Rio-Chanona, et al. (2021). This model neglects labor mobility, which is an important factor at the national level, but is less important for a large metropolis where workers seldom move. We are developing a new model that takes geographic mobility into account as well.

29 The links in the occupational transition agent-based model are based on the frequency of occupational job transitions, whereas our earlier job space paper was based on similarity of work activities.

30 Frey and Osborne (2013).

31 According to the Bureau of Labor Statistics a 'dispatcher' will 'Schedule and dispatch workers, work crews, equipment, or service vehicles for conveyance of materials, freight, or passengers, or for normal installation, service, or emergency repairs rendered outside the place of business. Duties may include using radio, telephone, or computer to transmit assignments and compiling statistics and reports on work progress.'

4: How Simulations Help us Understand the Economy

1 Steen, Lynn A. 1999. *Twenty Questions about Mathematical Reasoning. Developing Mathematical Reasoning in Grades K-12: NCTM Yearbook 1999.* Lee Stiff, Editor. Reston, VA: National Council of Teachers of Mathematics, 1999, pp. 270–285.

2 Another example is military simulation. See, for example, Hill and Tolk (2017).

3 https://tfresource.org/topics/TRANSIMS.html

4 See, for example, Duan et al. (2015), Davis et al. (2021) and Aleta et al. (2021). There were also very useful models for how the COVID virus behaves in the lungs, see Moses et al. (2021).

5 One problem with large agent-based models is that they can become black boxes, whose behavior is difficult to understand. This can be avoided by building models modularly, so that it is easy to experiment with them and identify what causes what.

6 Increasing returns provide a good example of the importance of non-linear behavior in economics. See for example Arthur (1994a).

7 When I moved to Los Alamos in 1981 I had the pleasure of being invited to Ulam's house several times. He made good margaritas.

8 Fermi et al. (1955); Stein and Ulam (1964).

9 Chinazzi et al. (2020).

10 Pangallo et al. (2023).

11 A synthetic population is a population of imaginary individuals whose characteristics match those of a real population – even if none of the individuals matches any particular individual in detail.

12 Geanakoplos et al. (2012). See also Axtell et al. (2014).

13 This is according to the Case-Shiller index: https://www.spglobal.com/spdji/en/index-family/indicators/sp-corelogic-case-shiller/sp-corelogic-case-shiller-composite/#overview.

14 Epstein and Axtell (1996).

15 Tax returns contain your name, age, income and address. By comparing tax returns in successive years, it is therefore possible to see how many people in each income and age group changed address, either leaving or entering a given area. (Someone inside the agency did this for us, so we never saw any names – only the probability of a person of a given age and income changing address.)

16 We used the Carroll (1997) permanent income model.

17 It is important to keep in mind that the housing model's outputs depend on all the model inputs along the way, such as interest rates, lending policy, etc. So, even though we ran thirteen years ahead of the starting point, the answers are somewhat anchored by the inputs. Only somewhat, though: Setting the parameters to sufficiently counterfactual values could dramatically change the outputs.

18 https://en.wikipedia.org/wiki/Case%E2%80%93Shiller_index

19 Baptista et al. (2016).

20 The central banks now using agent-based models, influenced by our early model for Washington DC, include Denmark (Cokayne, 2019), Ireland (Gurgone, 2020), Italy (Catapano et al., 2021), The Netherlands (Tarne et al., 2021), Spain (Carro and Stupariu, 2022) and Hungary (Mérő et al., 2022), as well as further work by the Bank of England (Carro et al., 2022).

Part II : Standard Economics ⇔ Complexity Economics

5: A Lightning Summary of Standard Economics

1 For more on utilitarianism, see https://plato.stanford.edu/entries/utilitarianism-history/.

2 Muth (1961).

3 Chevalier and Goolsebee (2009). This example was taken from Carlin and Soskice (2015).

4 Keynes (1936).

5 The account I present here, which is standard lore in economics, has been criticized as apocryphal by James Forder (2014). He argues that contemporary Keynesian economists were not as naïve as they were portrayed, and that they did not advocate using the Phillips curve as a justification to intentionally inflate the money supply, nor did this motivate policymakers.

6 Solow (1956).

7 Koopmans (1965) and Cass (1965).

8 Ramsey (1928).

9 Calvo (1983).

10 A standard model is Smets and Wouters (2007). See Glandon et al. (2022) for a review, and Edge and Gürkaynak (2011) for an evaluation of predictive performance.

11 Baqaee, Rezza and Farhi (2019). This model is not a micro-founded model, but rather builds on Leontief's original ideas. These models

incorporate the production network, and from this point of view have a complexity economics flavor.

12 Foster and Frijters (2012).

13 Smith (1962).

14 Knez and Smith (1987).

15 Gabaix (2020).

16 Gennaioli and Shleifer (2018).

17 In Cohen (2019), Naidu et al. prefaced the remark I cited by saying, 'In fact, all predictions and conclusions in economics are contingent: if x and y conditions hold, then z outcomes follow. The answer to almost any question in economics is "it depends", followed by an exegesis on what it depends on and why.' I wholeheartedly agree with this; economics should depend on conditions. However, that doesn't mean that we should be able to prove the efficacy of any policy recommendation we favor at the outset.

18 https://papers.rumsfeld.com/about/page/authors-note

19 Knight (1921).

6: Modeling an Uncertain and Complicated Economy

1 Simon, Herbert A. 1955. 'A Behavioral Model of Rational Choice'. *The Quarterly Journal of Economics*, vol. 69, no. 1, 1955, pp. 99–118. *JSTOR*, https://doi.org/10.2307/1884852.

2 I unfortunately never met Herbert Simon, who died in 2001. Given that he was clearly one of the pioneers of complex systems, and complexity economics in particular, it always struck me as peculiar that he never spent time at the Santa Fe Institute. I once asked SFI founder Murray Gell-Mann why this never happened. He answered that he had once served on a committee with Simon and felt intimidated by him. Given that Murray was the smartest and most intimidating person I knew (by a healthy margin), Simon must have been a real force of nature. For more on Herb Simon, see Axtell (2024).

3 Simon (1957), p.198.

4 Gigerenzer and Brighton (2009).

5 Gigerenzer and Brighton (2009). Note that heuristics can vary culturally, see Heinrich et al. (2004).
6 In statistics the maxim that simplicity is a virtue is called parsimony.
7 See Gigerenzer, Todd, and the ABC Research Group (1999).
8 Markowitz (1952).
9 Fama (1976).
10 The use of equal weights for portfolios is not the same as investing in an index. Instead, it means picking the assets one wants to invest in and investing equal amounts of money in each of them.
11 Despite the fact that it is now well-documented that sophisticated portfolio optimization methods based on Markowitz's original paper perform worse, out of sample, than equal weights, to my knowledge portfolio optimization is still taught in textbooks without any mention that it is useless (and in fact lowers performance). There are still highly profitable companies who sell this to clients, apparently with no awareness that simple heuristics are superior.
12 The term *bias–variance tradeoff* describes the problem of fitting a family of models with free parameters to data. The result can fail because the family of models is too simple, so that it does not contain the 'true' model. This is called *bias*. Or it can fail due to *variance*, which means that the family of models is too complicated, so that with limited data the resulting parameters overfit the data, and make poor predictions on data they have not seen. In my experience people reason about this poorly: Models are far more often wrong due to excessive variance than excessive bias.
13 As a result, heuristics can vary based on experience; see Bednar and Page (2007).
14 Gigerenzer and Selten (2001).
15 Vernon Smith (2008) independently proposed a concept he also calls ecological rationality, but he means something different, specifically how we solve problems collectively through developing institutions and social norms. Both are relevant.
16 Athey and Imbens (2019).
17 There is an important mainstream literature on replacing rational expectations with learning algorithms in traditional economic settings, pioneered by Thomas Sargent. See, for example, Sargent (1993).

18 See Arthur (1991, 1994b) for early attempts to design algorithmic agents that act like human agents and Epstein (2013) for 'Agent_Zero', which incorporates a cognitively plausible approach to human behavior in agent-based models.

19 Cars Hommes and his collaborators run economic experiments and explain the results of these experiments using simple agent-based models. See Hommes (2013).

20 Sims, Christopher A. 1980. 'Macroeconomics and Reality.' *Econometrica*, vol. 48, no. 1, 1980, pp. 1–48. *JSTOR*, https://doi.org/10.2307/1912017. Although the original quote concerned the wilderness of disequilibrium, this evolved into the wilderness of bounded rationality. Both are appropriate.

21 I should also say that not all complexity economists have entered through the back door and kept it in sight. Others have walked boldly into the interior, and let themselves get a bit lost while searching for better economic models. One problem with this is that models can become too complicated and hard to understand, and can be difficult to compare to data, but some of the models that have used this approach provide useful proofs of principle.

22 See Gode and Sunder (1993).

23 Although the concept of utility is useful for building conceptual models, it is limited as a basis for building fully quantitative models. The utility functions measured in experiments are complicated, even in simple settings (see Erev et al., 2017, or Plonsky et al., 2019). This makes it difficult to match reality quantitatively. To simplify the math, economists like to use simple functional forms for utility. For example, a common choice is to assume that utility is the logarithm of consumption, or consumption raised to a power – the choice is typically based on mathematical convenience, which means that the answers also depend on mathematical convenience rather than empirical evidence. Another problem is that utility is always assumed to be *maximized*. This means that the agents in the model don't just strive to make their utility bigger – all of them actually succeed in making it as big as it can be, given the situation. In simple situations this might be a reasonable assumption, but, as discussed in Chapter

7, it is unreasonable to expect that real agents could do this in more complicated situations.

24 Gabaix and Laibson (2008).

25 Finding the equilibrium for N assets is NP hard. Roughly speaking, unless there is an as yet unknown shortcut, this means that the effort required increases exponentially with the number of assets.

7: Why is the Economy Always Changing?

1 For examples where mainstream models produce endogenous business cycles, see Beaudry et al. (2015, 2016), Day (1981, 1982) and Boldrin and Montrucchio (1986). Blume and Easley (1991) show how endogenous dynamics can arise in a more general context.

2 For other interesting examples of emergent behavior, see, for example, Gualdi et al. (2015) or Dessertaine et al. (2022).

3 See, for example, Veblen (1899).

4 Asano et al. (2021).

5 When it copies the rate of its best neighbor, each household makes a small random error, so that its new savings rate is a bit different from the one it copied. This prevents the economy from getting stuck in exactly periodic cycles.

6 This is particularly surprising given that the social-interaction time is effectively a measure of the inattention of the households. A large social-interaction time means that households only occasionally tune into what the other households are doing – most of the time they just ignore the world and keep on doing what they have been doing. It is therefore surprising that households become richer by paying less attention. The reason this happens is because longer interaction times allow more effective selection – it means that the household that is copied has kept its savings rate constant longer, and that it has had more time to reap the benefits of saving.

7 Dosi et al. (2010), and Dawid et al. (2012).

8 Scheinkman (1976).

9 Nash (1951).

10 In fact, chaos in rock-paper-scissors only happens if the payoffs are asymmetric, meaning that one of the players has an inherent advantage over the other. In other words, if Bob wins he gets more points than Susie does when she wins.

11 Sato, Akiyama and Farmer (2002).

12 See also https://www.tiktok.com/@rockpaperscissorsbattle/

13 This makes rock-paper-scissors particularly unstable and is part of the reason why the learning dynamics do not converge to equilibrium even though it is a very simple game.

14 Galla and Farmer (2013).

15 Studying games with an infinite number of moves might seem like a strange choice – why not start with the simplest case, like games with only two possible moves? In physics it is often the case that understanding the behavior of an infinite number of particles is simpler than understanding a finite number (greater than two). We were building on some remarkable work by the German physicists Sigurd Diederich and Manfred Opper, who studied random games with an infinite number of moves, and showed analytically how the number of Nash equilibria depends on the degree of competitiveness. See Diederich and Opper (1989) and Opper and Diederich (1992).

16 Pangallo et al. (2022). See also Sanders et al. (2018).

17 Pangallo, Heinrich and Farmer (2019).

Part III : The Financial System

8: Inefficient Markets

1 Buffet, Warren E. 1988. *Berkshire Hathaway Letters to Shareholders 1965–2014*. Max Olson, Editor. Sanage Publishing House, 2015.

2 The number of active contracts is the number of people in the world – roughly 8 billion – times the number of contracts per person, plus all the contracts that exist between businesses, and businesses and governments. I estimate that the typical person in the developed world has many thousands of active contracts. A trillion total contracts is

probably a low estimate. When we include businesses and governments there are likely tens or hundreds of trillions of contracts in existence at any point in time.

3 As in physics. People often say that physics has had a big influence on economics, but the influences also go the other way. Joule, who formulated the conservation of energy, was an accountant for a brewery. See Mirowski (1989).

4 Greenwood and Scharfstein (2013).

5 Philippon and Reshef (2009).

6 Goldin and Katz (2008).

7 See also the ECB conference on 'The Optimal Size of the Financial Sector', https://www.ecb.europa.eu/events/conferences/html/140902_opt_size.en.html and Coeure (2014).

8 How we came to be able to predict sunspots is a fascinating story. In the seventeenth century, using the newly invented telescope, monks in Belgium began making daily recordings of sunspot activity. They handed this off to future generations in the form of a long record of sunspot activity. In 1927 the British polymath Udney Yule invented the concept of the time-series model and showed that it could be used to predict sunspot cycles.

9 Farmer and Sidorowich (1987).

10 Fama (1991).

11 Bachelier (1900).

12 The ratio of annual returns to annual risk is called the Sharpe ratio. The Sharpe ratio of the S&P index is about 0.5; for Prediction Company it was about 2.5. This means that our profits were very steady, far steadier than proponents of efficient markets deemed possible.

13 I have been unable to find the original source, so this is based on my memory.

14 I left Prediction Company to return to academia, according to my original plan, in 1999. Prediction Company was sold to UBS in 2006. The Dodd-Frank Bill implemented the Volcker Rule, which prohibited investment banks from trading for themselves and for their clients at the same time, and as a result the company was sold to the Millennium stable of hedge funds in 2013. Despite remaining profitable, it

was recently disbanded by Millennium, for reasons that are not appropriate to explain here.

15 For this to be true markets need to be 'complete', which means that there are forums where the underlying assets can be bought and sold so that prices can adjust accordingly. In addition, this requires the assumption of a rational-expectations equilibrium.

16 Cabannes et al. (2018).

17 See, for example, Shiller (1981) and Summers (1986).

18 The airline company would of course prefer to buy an option for jet fuel, but this isn't possible. However, since the price of jet fuel is closely correlated to the price of oil, if the price of oil goes up above the strike price of their option, they can sell the option for a profit and use this to offset their higher fuel costs.

19 Thorp and Kassouf (1967); Black and Scholes (1973).

20 The US SEC requires mutual funds to report their positions every quarter. Warren Buffett was once granted a waiver from doing this in a timely manner. He was in the process of buying 5 per cent of IBM, and he wanted nine months so that he could do this gradually and in secrecy, since if the information that he was buying leaked out, others would buy the stock too and drive the price up before he could take his position.

21 The square-root function is infinitely steep at the origin, so a purchase of a single share has a far larger impact per share than a purchase of a million shares. The square-root law is well supported by the data, as long as there is an 'orderly market'. There are circumstances, such as financial crises, where the prediction is violated, but under normal conditions it works very well.

22 For historical reasons, economists usually sketch price as a function of supply or demand rather than the other way around.

23 Stanley et al. (1996); Mantegna and Stanley (1999).

24 For a thoughtful reflection on data's relationship to modeling, see Page (2018).

25 While Descartes had a theory that purportedly explained *why* gravity existed – which Newton did not – Newton's theory was far more useful. See Farmer (2013) for a discussion of how physics mainly uses inductive logic, whereas economics mostly uses deductive logic.

26 Even if we assume market clearing, when the supply or demand functions change, the price will move, and we will observe market impact. So, in this case, market impact is a description of how the equilibrium has changed. But if prices do not immediately move to a new equilibrium, market impact tells us about the rate of movement. In either case, it is a very useful thing to know.
27 Bak, Paczuski and Shubik (1997) is one example of earlier work that assumed random order placement.
28 In our theory, buy orders were placed randomly at all prices below the current best bid, and sell orders were placed randomly at all prices above the current best offer.
29 Daniels et al. (2003).
30 The zero-intelligence assumption was pioneered by Gode and Sunder (1993). See Smith et al. (2002).
31 Farmer, Patelli and Zovko (2005).
32 Bouchaud et al. (2018).
33 Gode and Sunder (1993).
34 Grossman and Stiglitz (1980).
35 Farmer (2001).
36 In the following decade, the value of this signal decreased substantially, but it nonetheless remained valuable for some time after we evaluated it in 1998.
37 Farmer (2002). See also Summers (1986), which presents some related arguments.
38 Cherkashin et al. (2009).

9: The Self-Referential Market

1 Twain, Mark. 1894. *Pudd'nhead Wilson*. London: Chatto and Windus, 1894, p. 130.
2 Carlson (2007).
3 The point that speculative feedback strategies such as trend-following can be destabilizing to markets has also been made by Cutler, Poterba and Summers (1991).

4 You may wonder how the market ever recovers from such feedback loops. Eventually, it becomes so obvious that stocks are dramatically underpriced that enough courageous investors step in and start buying to stop the fall; if their timing is good, they make profits.

5 Clustered volatility in stock prices has two immediate causes. One is that trading volume varies; the other is that the size of the price-response to each trade varies. In a 'liquid' market, the size of price-responses to trades is small. We showed that liquidity is a more important determinant of price movements than trading volume. (Gillemot et al., 2007.)

6 Cutler, Poterba and Summers (1989).

7 Several other studies have been done since then, and have found similar results. A comprehensive study done on shorter timescales, which compared prices to a newsfeed, found that on short timescales an even smaller fraction of price changes are associated with information arrival. See Joulin et al. (2008).

8 Black (1986). His actual statement was, '[. . .] we might define an efficient market as one in which price is within a factor of 2 of value, i.e., the price is more than half of value and less than twice value. The factor of 2 is arbitrary, of course. Intuitively, though, it seems reasonable to me, in the light of sources of uncertainty about value and the strength of the forces tending to cause price to return to value. By this definition, I think almost all markets are efficient almost all of the time. "Almost all" means at least 90%.'

9 Shiller (1981).

10 See, for example, Palmer et al. (1994), Arthur et al. (1996).

11 In a classifier system, each agent is given a set of rules. Each rule consists of an input condition and an output if that condition is met. Successful rules reproduce, albeit imperfectly, with modifications to the rules via random mutations and recombination, and unsuccessful rules are removed from the population, so that the system learns over time.

12 Brock and Hommes (1997).

13 Many in the cast of characters I have described above have gone on to play significant roles in complexity economics. Brian Arthur, a master storyteller who speaks with a soft and gentle Irish brogue, was one of

the first to illuminate the unstable and nonintuitive behavior of economies with increasing returns. He later wrote a very influential paper called 'The El Farol Problem', where the discussion is anchored around a popular bar down the street from the first home of the Santa Fe Institute, and coined the term 'complexity economics'. Blake LeBaron became a leading expert on agent-based modeling in finance, and has provided a great deal of insight into the origins of booms and busts. Cars Hommes went on to start a large group at the University of Amsterdam, where he has performed many experiments that illuminate when markets converge or do not converge to equilibrium, explaining the behavior of the subjects of his experiments with simple agent-based models.

14 Platt (2022).

10: Ecology and Evolution in Finance

1 Understanding hard problems can take years, decades, centuries or even millennia. Markets and financial strategies have been evolving since Mesopotamian times. This is not to underestimate the Mesopotamians! Herodotus describes sophisticated Babylonian auctions, and the first known recorded commercial laws in Babylonia banned monopolies and included definitions of the different types of monopolies. Such laws were not introduced in the United States until the 1890s.

2 Farmer (2002), Farmer and Skouras (2013) and Scholl et al. (2021).

3 See Cutler, Poterba and Summers (1991).

4 Fama and French (1996). The way in which trend-following strategies can survive in the market was shown by Delong et al. (1990).

5 Cornering a market happens when a single player owns enough of the market to have monopoly power, so that it can set prices to its own advantage. This is a good example where markets need to be regulated.

6 See Lotka (1925) and Volterra (1926). Analogues of the Lotka-Volterra equations have also been used by Paul Samuelson (1974) and Richard

Goodwin (1967). Their use of them was ad hoc – they simply postulated that such equations might apply. In Farmer (2002), I showed how these equations can be derived from a simple set of plausible assumptions about market impact and capital flows.

7 Farmer and Joshi (2002).

8 See, for example, Dosi and Nelson (1994).

9 Farmer and Lo (1999).

10 Lo (2004, 2019).

11 Levin and Lo (2021).

12 Scholl et al. (2021).

13 In ecology, it is also common that the properties of an ecosystem depend on the relative populations of the species in it. This is called density dependence. In Scholl et al. (2021), we showed that financial ecosystems are strongly density-dependent.

14 A particularly interesting case occurs when the market is efficient. As an approximation, we neglected risk and defined efficiency as the situation where the returns of all strategies are equal. If we artificially adjusted the wealths of the three strategies to make the market perfectly efficient, all three strategies had a mutualistic relationship – that is, increasing the wealth of any of them increased the returns of the others.

15 Zhou (2021).

16 Darley and Outkin (2007).

17 There is a close parallel in the downfall, in 1997, of the high-flying firm LTCM, as described in Roger Lowenstein's book *When Genius Failed*. LTCM had large, highly leveraged convergence bets on a variety of international bond spreads. When they started to unwind their huge positions, they caused bond spreads to widen in unison, threatening the stability of the entire global bond market.

18 Khandhani and Lo (2011). Their analysis indicates this was only a four-day event, but I think that is because they did the analysis with an extremely simplified stat-arb strategy. Prediction Company's losses suggested that liquidation activity occurred for more than ten days in late July and early August.

19 In fact, with sufficiently high leverage, a single fund liquidating its positions can cause itself to become bankrupt. See Caccioli, Bouchaud

and Farmer (2012), where we show how this can happen and propose a valuation method that would make markets more stable.

11: How Credit Causes Financial Turbulence

1 Polanyi, Karl, 1886–1964. *The great transformation: the political and economic origins of our time*. Foreword by Joseph E. Stiglitz; with a new introd. by Fred Block.—2nd Beacon Paperback ed. p. cm. Originally published: New York: Farrar & Rinehart, 1944 and reprinted in 1957 by Beacon in Boston.
2 Reinhart and Rogoff (2009).
3 Another important amplifier of systemic risk was the loss of faith in the banking system. There was effectively a bank run on the shadow banking system that removed its ability to provide credit. See Summers (2000) for a review.
4 In this example I assume that all the savings of the owner are in the equity in her house. It is often convenient to think about leverage one investment at a time, but for most purposes it is more relevant to consider all your equity and all your assets. For example, if, in addition to your \$500,000 house with your \$400,000 mortgage, you have \$300,000 in the bank, then your total assets are worth \$800,000 and your total equity is \$400,000, so your total leverage is 2:1.
5 As shown in Caccioli, Bouchaud and Farmer (2012), this is a good example where the market impact of unwinding a large position needs to be taken into account in valuing a portfolio. Even without any external events, LTCM's own market impact might easily have driven them bankrupt if they had been forced to rapidly liquidate their portfolio.
6 John's theory of the leverage cycle built on a surprising result that a single equation for supply and demand could determine both interest rates and leverage, which was published in the 1997 SFI conference 'The Economy as an Evolving Complex System II', which was a follow-up to the original conference discussed in Chapter 9 (Geanakoplos, 1997). The full theory was published in Geanakoplos (2003).

See also Fostel and Geanakoplos (2008) and Geanakoplos (2010). The fact that leverage can amplify crashes was discussed earlier in Gennotte and Leland (1990).

7 The assumptions include no taxes, no bankruptcy costs, no agency costs and no asymmetric formation, as well as efficient markets.

8 This already violates the widely held interpretation of the Modigliani-Miller theorem that prices should not be affected.

9 I am abbreviating the story of this collaboration. Originally we were hoping to simulate credit networks involving multiple assets in collaboration with Duncan Watts and Alan Kirman. This proved to be too difficult because of the challenges of doing market clearing on multiple assets, so we simplified the problem to study a single asset.

10 Thurner, Farmer and Geanakoplos (2012).

11 When funds defaulted, we created a new fund with the same characteristics and a modest amount of capital. In most cases, the fund's assets would grow over time, and if we ran long enough, there would eventually be another big crash and they would default again.

12 During a taxi ride from LaGuardia to midtown Manhattan in 1998, my driver talked non-stop about the great tech funds I should invest in. Thinking this indicated that we were close to the peak of the bubble, I immediately sold half my stock portfolio. This might sound like a great move, but in the next two years the market went up by another factor of 2 before it reached its peak and came tumbling down. Fortunately, the fall was gradual and I managed to unwind the rest of the portfolio while prices were still fairly high.

13 Poledna et al. (2014).

14 The drying up of credit during the 2008 financial crisis caused a bank run on the shadow banking system, which further stressed institutions that depended on credit. The causes of bank runs were studied in Diamond and Dybvig (1983).

15 https://www.bis.org/publ/bcbs04a.htm

16 https://www.bis.org/publ/bcbs107.html

17 Danielsson et al. (2001).

18 From Aymanns et al. (2016).

19 From Aymanns and Farmer (2015).

20 Aymanns and Farmer (2015).

21 Aymanns et al. (2016).

22 The relative wealth of the bank vs the fund has essentially the same effect as another parameter that controls the risk appetite of the bank. When the risk appetite is large, the bank uses higher leverage. Taking more leverage allows the bank to buy more of the asset.

23 The critical leverage depends on the overall volatility of the asset. The huge leverages of 30 or 40 used by investment banks are for carefully hedged portfolios that have very low, unleveraged volatility. While it's easily possible to estimate the critical leverage for the financial system, doing so accurately requires a more detailed model.

24 Basel II was not officially adopted by the Basel Committee until 2004, but VaR was widely used before then.

25 Farmer et al. (2020). See Aymanns et al. (2018) for a review.

26 Wiersema et al. (2023).

27 Ghashghaie et al. (1996).

28 It is well known that aggregating variables with fat tails makes the tails thinner, and there are precise mathematical rules for how this occurs. Nonetheless, the resemblance of the clustered volatility in both cases, and the quantitative match, is striking.

29 Muzy et al. (2000).

30 Calvet and Fisher (2001).

31 In 1996, when Ghashghaie's study was done, the fastest trading strategies had reaction times on the order of a hundredth to a thousandth of a second. Now the reaction time can be as low as 50 microseconds, with turnaround times that are themselves a fraction of a second.

Part IV : Climate Economics

1 See IPCC (2018), https://www.ipcc.ch/sr15/

12: How we Learned to Predict Weather and Climate

1 Lorenz, Edward N. 1972. 'Predictability: Does the flap of a butterfly's wings in Brazil set off a tornado in Texas?' Presented before the American Association for the Advancement of Science, December 29, 1972. https://eapsweb.mit.edu/sites/default/files/Butterfly_1972.pdf

2 Much of the account given here is drawn from Edwards (2010), which I highly recommend.

3 https://www.weather.gov/timeline

4 The Norwegian Vilhelm Bjerknes also suggested the idea of numerical weather forecasting, independently of Abbe. His son Jakob developed the theory of El Niño.

5 Richardson (1922).

6 Charney, Fjörtoft and von Neumann (1950).

7 Lazo, Morss and Demuth (2009).

8 Bauer, Thorpe and Brunet (2015).

9 National Research Council (1979).

10 IPCC (2014), https://www.ipcc.ch/report/ar5/syr/

11 This might bring to mind the Borges story 'On Exactitude in Science', written in 1946, in which he imagines an empire where the science of cartography becomes so exact that only a map on the same scale as the empire itself will suffice. This is not what we have in mind! Any simulation is only a caricature of the real thing; even if there is a simulated agent corresponding to every human on earth, the simulated agent is at best a pale reflection of the real agent, containing vastly less information. Models are, by definition, highly compressed renderings of reality. See Rissanen (2007).

12 A skeptic might argue that weather simulations are based on hard-and-fast laws of physics, whereas there are no such laws in economics. This is partially true – some aspects of weather depend on well-understood phenomena, like fluid flow. However, many essential aspects of weather, such as cloud formation, are not fully understood from first principles, and require heuristics for their solution. While I can't prove this yet, my intuition is that the instabilities of the economy are weaker than those of the weather, the measurement problems

for the economy are easier, and the simulation power required is far less.

13: Climate Economics and Technological Progress

1 Nordhaus, W. D. and Boyer, J. G. 2000. *Warming the World Economic: Models of Global Warming*, Cambridge, Massachusetts: The MIT Press, p. 69.
2 Nordhaus (2017).
3 Stern (2006), Stiglitz (2006), Sachs et al. (2019) and Stern et al. (2021).
4 The fact that services have gone up in price and high technologies have dropped in price is related to the Baumol effect, which is at least partially explained by our model for technological progress discussed in Chapter 3.
5 Moore (1965). Transistors alone have dropped in cost at a rate of about 40 per cent per year for the last seventy years. They are now so small that they are approaching the molecular limit, and Moore's Law as a scaling law for transistor density must break down. It will be interesting to see how this affects other aspects of the rate of progress in computers, such as their performance and cost. Once we can't make the components of integrated circuits any smaller, will technological progress in computers grind to a halt? In any case, the rate is likely to change.
6 Koh and Magee (2006, 2008).
7 Wright (1936).
8 There are usually multiple criteria for measuring the performance of a technology, so stating a single rate of improvement can be difficult. Consider energy generation: Power is measured in watts, and so it might seem like a watt is a watt. But there are clean watts and dirty watts, steady watts and intermittent watts, and safe watts and unsafe watts. Nuclear power, for example, went up in price at least in part because of changing safety standards.
9 McNerney et al. (2011).
10 Farmer and Makhijani (2010).
11 *The Economist* (2014).

12 These numbers do not quite agree with those in Farmer & Makhijani. This is because there we used US numbers, whereas here, to be consistent with Way et al. (2022), I am using global averages.
13 The Performance Curve Database can be found at pcdb.santafe.edu.
14 Nagy et al. (2013).
15 See, for example, Sahal (1982).
16 Farmer and Lafond (2016).
17 Lafond et al. (2018).
18 Lafond, Greenwald and Farmer (2020).
19 Way et al. (2022).
20 Useful energy measures the work that is extracted from each energy source. Electricity-generation technology costs are levelized costs of electricity (LCOEs), which take factors such as amortization and the cost of capital into account. Battery series show capital-cost per cycle and energy stored per year, assuming daily cycling for ten years (these are not directly comparable with other data series here). Modeled costs of power-to-X (P2X) fuels, such as hydrogen or ammonia, assume historical electrolyzer costs and a fifty-fifty mix of solar and wind electricity. See Way et al. (2022).
21 There is some evidence that the prices of oil and gas increased around 1980 and have never returned to the same level, but this is still small compared to the transformations in renewable energy prices.
22 Burning ammonia can produce NOx (smog) if it isn't burned cleanly, but there are good ways to prevent the smog from developing.
23 The most remarkable cost declines for energy-storage technologies are for capacitors, which have dropped in cost and increased in performance at rates comparable to computers. Extrapolation suggests that they could become cost- and performance-competitive with batteries circa 2030. This would be a major breakthrough, because capacitors are solid-state devices that can be more reliably operated and maintained than batteries and can be charged and discharged quickly. See Koh and Magee (2008).
24 Meng et al. (2021).
25 Way et al. (2022).

26 While *costs* of improving technologies like solar photovoltaics or wind tend to follow Moore's Law, their *deployment* follows S-curves, meaning that it initially increases exponentially and then eventually levels off.

27 Stern (2006).

28 There are proposals afoot to develop small modular nuclear reactors that can be mass-produced. Conceivably this will lower the costs, but it runs up against fundamental economies of scale. The probability of success is difficult to estimate, and it is unlikely that this will happen quickly. Modular nuclear reactors would be a new product, and only rarely are new products developed and brought to market at sufficient scale to drive substantial price declines in only a decade. There is no good evidence that this will result in lower costs. It is worth reiterating that (after adjustment for inflation) most technologies have historically declined in cost at very slow rates – computer technologies and the renewable energy technologies mentioned are special.

29 The real costs of nuclear power are much higher than the stated costs. The two main reasons for this are insurance and waste disposal. Because insurance companies refuse to insure the full risk of nuclear power plants, it is borne by the public. Very little nuclear waste has yet been disposed of, and this remains a work in progress. It is difficult to make good cost estimates, but these two factors substantially inflate the costs. We have neglected these costs here.

30 Farmer et al. (2019) discusses the idea of finding 'sensitive intervention points' to accelerate the transition.

Part V: Modeling for a Better Future

14: Guidance to Solve Some of Our Big Problems

1 Dosi et al. (2010), delli Gatti et al. (2011), Dawid et al. (2012), Lamperti et al. (2018).

2 Poledna et al. (2022).

3 Hommes et al. (2023).

4 Zucman and Saez (2019).
5 Turchin (2023).
6 See, for example, Piketty (2015), Piketty et al. (2018).
7 The agent-based models of Dawid et al. (2012) and Lamperti et al. (2018) have produced good qualitative insight into the origins of inequality; we now need to do this more quantitatively.
8 See Maarten Scholl's Oxford DPhil thesis (2023).
9 Properly estimating carbon emissions requires a complete knowledge of the production network. This is because it is necessary to take the emissions of the inputs to a product into account, and most firms have many inputs, and each of these inputs has many inputs. It only takes a few levels to reach a large fraction of the economy.
10 To see how this can be done, see Pichler, Lafond and Farmer (2020).
11 This module incorporates a set of ideas and methods developed by Cesar Hidalgo, now at the University of Toulouse, and my Santa Fe Institute colleague Ricardo Hausmann, who is also a professor at the Harvard Kennedy School. Hidalgo and Hausmann (2009).
12 Jackson (2016).

15: Removing the Roadblocks

1 https://en.as.com/latest_news/how-much-money-did-cerns-large-hadron-collider-cost-to-build-and-who-paid-for-it-n/
2 Mungo et al. (2022).
3 The idea of fitting parameters to data was first demonstrated by Adrien-Marie Legendre in 1806. Legendre showed how to find the parameters of a line, which has parameters for the slope and the *y*-intercept, so that the sum of the squared distance between the data points and the line are as small as possible. Legendre (1806), with the least-squares method appearing in the appendix.
4 Platt (2022); Kolic et al. (2022).
5 To give another example from physics, Ed Witten, perhaps the most brilliant living theoretical physicist and the unquestioned leader of string theory, has received numerous awards and honors, including

the Fields Medal and the National Medal of Science. Hundreds of the best physicists have been working on string theory for the last thirty years and have been strongly influenced by Witten, and no one questions his brilliance. But unless the predictions of string theory are confirmed in experiments, his work will not win a Nobel Prize.

16: Becoming a Conscious Civilization

1 Eisenhower, Dwight. *Diary*, 7/2/53, PDDE, 14: 358–60.
2 For some preliminary ideas for how this might be done see Bednar and Page (2007).

Index

Page references in *italics* indicate images.